# ARMCHAIR CHEMISTRY

ARMCHAIR
GUIDES

# ARMCHAIR
# CHEMISTRY

### EVERYTHING YOU NEED TO KNOW, FROM CATALYSTS TO POLYMERS

JOEL LEVY

Inspiring | Educating | Creating | Entertaining

Brimming with creative inspiration, how-to projects, and useful
information to enrich your everyday life, Quarto Knows is a favorite
destination for those pursuing their interests and passions. Visit our
site and dig deeper with our books into your area of interest:
Quarto Creates, Quarto Cooks, Quarto Homes, Quarto Lives,
Quarto Drives, Quarto Explores, Quarto Gifts, or Quarto Kids.

© 2010 Quarto Publishing plc

This edition published in 2018 by Chartwell Books,
an imprint of The Quarto Group,
142 West 36th Street, 4th Floor,
New York, NY 10018, USA
**T** (212) 779-4972 **F** (212) 779-6058
**www.QuartoKnows.com**

Conceived, designed and produced by Quid Publishing,
an imprint of The Quarto Group,
The Old Brewery,
6 Blundell Street,
London N7 9BH
United Kingdom

Chartwell Books titles are also available at discount for retail,
wholesale, promotional, and bulk purchase. For details, contact
the Special Sales Manager by email at specialsales@quarto.com
or by mail at The Quarto Group, Attn: Special Sales Manager, 401
Second Avenue North, Suite 310, Minneapolis, MN 55401, USA.

10 9 8 7 6 5 4 3 2 1

ISBN: 978-0-7858-3596-7

Printed in China

# CONTENTS

# CHEMISTRY: AN INTRODUCTION

Technically speaking, chemistry is the study of the elements and the compounds they form, but in its broader sense it is so much more—the science of everyday life, of the matter that makes up the world and how every single one of us can transform it in ways that seemed magical to our ancestors, and can still seem fantastic today.

For many readers "chemistry" will summon up images of test tubes and Bunsen burners, white lab coats, strange smells and the vague hope of an explosion. This fleeting, school-days acquaintance with the subject does it little justice, as this book sets out to prove. In these pages you will learn how chemistry transformed mankind, gave life to civilization, captured the imagination of mystics and magicians, and inspired the greatest minds in history. You won't need any prior knowledge of the subject, as everything from the most basic concepts to the most profound laws of matter will be explained in clear, concise fashion. You will need to bring along your sense of adventure and your capacity for wonder. Along the way you will meet some strange and incredible characters and learn a host of fascinating trivia, from why bread rises and ice floats to how a lizard can walk on water.

## Atoms, Elements, Molecules, and Compounds

Before going any further, we need to introduce some very basic terms and concepts. Two pairs of terms that chemists use a lot are "atoms" and "molecules," and "elements" and "compounds." What's the difference between them? An atom is the smallest unit of any substance. A molecule is two or more atoms joined together by chemical bonds. An element is the purest form of matter, one which is not made up of any other ingredients. At the moment there are 118 known elements; any substance that is not a pure element is made up of a combination of elements. Such a combination is called a compound. An element is made up of atoms of that element, and all the atoms of an element are identical to all the others. For instance, every atom of the element carbon is identical to every other carbon atom. For some elements

"Chemistry . . . offers one of the most powerful means towards the attainment of a higher mental cultivation . . . because it furnishes us with insight into those wonders of creation which immediately surround us."—*Justus von Liebig*

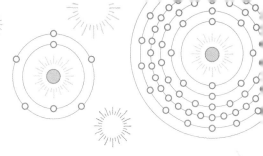

the atoms join up with each other to make small molecules. Oxygen, for instance, in its pure form, is a gas made up of molecules, each of which is composed of two oxygen atoms bonded to one another. Compounds and the molecules they are made up of are formed during chemical reactions, where atoms and molecules interact and adopt new combinations and arrangements. For instance, when the elements carbon and oxygen react to form the compound carbon dioxide, carbon atoms react with oxygen molecules to form molecules of carbon dioxide. All of these terms are explained in more detail in the book (for atoms, see pp. 26–27; for elements, see pp. 22–23; for chemical bonds, see pp. 78–79; for chemical reactions, see pp. 40–41).

## Quest for the Philosopher's Stone

Chemical knowledge is growing all the time, faster than most people imagine. More than 8 million different chemical substances, both naturally occurring and artificial, are now known; as recently as 1965 only 500,000 had been characterized and produced, and even this far outstripped the wildest imaginings of the chemists of 200 years ago. To avoid being overwhelmed by the sheer breadth of the subject, this book focuses on inorganic chemistry, the branch that deals with all the elements other than carbon, and with their compounds. Some simple carbon compounds, like carbon dioxide and calcium carbonate (aka chalk or limestone), also fall within the remit of inorganic chemistry.

Most of the history of chemistry has been concerned with inorganic chemistry, so their stories are largely one and the same—and what a story it is. One of the great adventures in the history of ideas, the development of chemistry is a tale of obsession, greed, danger, hope, and inspiration; each chapter of this book relates a chapter of that tale. Following the millennia-long quest to uncover the secrets of matter, it charts mankind's earliest steps in chemical manipulation of the environment with fire and cooking; the development of alchemy, that mixture of science and magic concerned with the Philosopher's Stone, the Elixir of Life, the Scientific Revolution, and the search for the elements. Building to a crescendo with the discovery of periodic law, the key to all of chemistry, the book ends with the fulfillment of an age-old quest: the transmutation of elements.

At every stage in this epic intellectual journey, the essential concepts of chemistry are introduced and explained in accessible style with a minimum of mathematics and formulae. Step-by-step exercises help you to grasp and master the principles of chemistry so that you too can decipher the hidden language of the elements that makes it possible to read the book of nature.

Chapter

1

# Chemistry in the Ancient World

This chapter explores the rich chemical heritage
of antiquity and beyond, reaching back in time to
the prehistoric era, introducing along the way some
of the fundamental concepts of matter and energy,
not to mention the secret of good toast. Pre- and
ancient history may predate science but they were
eras of great technological sophistication, including
surprisingly advanced chemical technologies, and
they also witnessed the genesis of modern
conceptions of matter and the elements.

# PREHISTORIC CHEMISTRY

Chemistry seems a quintessentially modern science; indeed, during the Enlightenment it was regarded as *the* science, lighting the way out of the Dark Ages and into the brave new world of the laboratory. But in fact chemistry is as old as human culture—you could say it was our use of chemistry that made us human. Whether they realized it or not, prehistoric chemists dating back to the dawn of mankind were employing the basic principles of chemistry.

## Fire, Work with Me

One of the turning points in human evolution was when our ancestors first began to exert control over their environment, using the principles of combustion. Combustion is the layman's term for the oxidation of carbon: the bonding of carbon with oxygen, in an exothermic chemical reaction (one that releases energy in the form of light and heat)—in other words, fire. The conditions for spontaneous combustion of carbon are very rare on Earth because typical combustion reactions require

activation energy—an energy boost that gets the reaction started (see pp. 46–47).

There is evidence that *Homo erectus*, one of the species ancestral to modern humans (*H. sapiens*), used fire to clear habitats and probably to flush out game. Possibly the earliest fire-users took advantage of naturally occurring fires started by lightning strikes, but by the time *H. sapiens* had evolved, if not before, our ancestors had learned how to generate enough activation energy to start combustion for themselves (either by striking sparks from flints or by rubbing bits of wood together to generate heat by friction).

A host of other technological advances followed on from this, involving a range of chemical technologies. First, and perhaps most important for human evolution, was the discovery of the chemistry of cooking (see pp. 14–15), which made edible a much wider range of food and opened up new dietary possibilities with high calorie and protein content.

# THE AGES OF METAL

The chemistry of pyrotechnology made possible the development of metallurgy, and the development of prehistoric civilization from the Stone Age, to the Copper Age, to the Bronze Age, and finally to the Iron Age. The sequence of these Ages was determined by the chemical properties of the metals—their propensity to react with oxygen and other elements in turn determines how commonly they occur in the environment in their pure forms. Pure metals are easier to find, recognize, mine and work. The least reactive metal is gold, which was probably the first metal worked by humans, but it is a soft, malleable element which had little utility to prehistoric man beyond ornament. Also resistant to oxidation, and therefore found in its pure form in nature, is copper. Initially copper was worked without heating, but pyrotechnology allowed ores to be smelted and metals to be melted and cast. Ores of copper and tin are sometimes found together, and when smelted they would have produced an alloy—a mixture of metals, in this case bronze. Iron ores are more abundant than those of copper and tin, but iron has a much higher melting point, making it hard to smelt until kiln technology had advanced. By around 1100 BCE, however, ancient metallurgists discovered that reheating impure iron with charcoal produced steel, which gets its strength and ability to take a sharp edge from the addition of carbon.

## Pyrotechnology

Mastery of the chemistry of combustion led to many other "pyrotechnic technologies," each directly involving chemistry. One of the earliest was the treatment of the pigment ochre. Ochre is the common name for a type of clay tinted yellowish-brown with the iron ore haematite—a compound of iron known as hydrated iron(III) oxide, which has the chemical formula $Fe_2O_3$ (see pp. 136–137 for an introduction to chemical formulae). Prehistoric man discovered that heating ocher at 500–540°F (260–280°C) caused a chemical reaction known as calcination, which produced a still wider range of colors—in particular a striking red.

According to the paleoanthropologist Richard Rudgley, "The pyrotechnic arts were first applied to ocher and this technology may have then been transferred to other materials such as flint and clay." Heat treatment changed the crystalline structure of flint to produce sharper edges and better tools, while firing clay gave birth to the craft of pottery.

The leading scholar of ancient "pyrotechnology," Theodore Wertime, argued that prehistoric man could be considered a sophisticated chemist: "Stone Age men were using fire in manifold versatile ways from 25,000 years ago . . . from the tempering of wood spear points, the oxidation of such pigments as ochre, the annealing of stone projectile points, the fire setting of quarries (from which the lime kiln and possibly the metallurgic furnace emerged), to the development of the technologies of the cooking hearth."

# 1 Fermentation of Ethanol

## THE PROBLEM:

The process of "brewing" to produce alcoholic drinks by fermentation of fruit and grains is over 5,000 years old. Nowadays, ethanol (ethyl alcohol) production by fermentation of sugars for use in beverages, as a solvent in industry and as a fuel is very important industrially. Starting from one kilogram of pure glucose (also called dextrose), how can we determine how much ethanol would be produced by the process of fermentation?

## THE METHOD:

Generally, ethanol may be produced by the fermentation of simple sugars such as glucose, fructose, and sucrose. Fermentation is essentially an anaerobic biological process (meaning that it takes place in the absence of air) in which a series of chemical reactions convert the sugars to ethanol in the presence of microbes such as yeast (*Saccharomyces cerevisiae*).

The simplified equation for the fermentation of the simple sugar, glucose (dextrose), is:

$$C_6 H_{12} O_6 \longrightarrow 2\ CH_3 CH_2 OH + 2\ CO_2$$

$$\text{glucose} \qquad \text{ethanol} \qquad \text{carbon} \\ \text{dioxide}$$

The equation shows that 1 glucose molecule is converted into 2 ethanol molecules and 2 molecules of carbon dioxide gas.

To solve the problem, we need to calculate the relative molecular masses ($M_r$) of each molecule. To do this, we must first know the atomic weight ($A_r$) of the atoms in each molecule. They are: carbon (C = 12), hydrogen (H = 1) and oxygen (O = 16). (For more on how to calculate the atomic weight of elements, see Exercise #7 on p. 62.) Once we have calculated the relative molecular masses and plugged these into the equation on page 12, we will be able to work out the amount of ethanol produced from the fermentation of 1 kilogram of glucose.

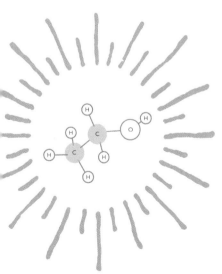

## THE SOLUTION:

Glucose ($C_6 H_{12} O_6$) contains 6 x carbon, 12 x hydrogen and 6 x oxygen atoms. The relative molecular mass of glucose = (6 x 12) + (12 x 1) + (6 x 16) = 180. Similarly, for ethanol ($CH_3 CH_2 OH$ or, as it is commonly written, $C_2H_5OH$), $M_r$ = (2 x 12) + (5 x 1) + (1 x 16) + (1 x 1) = 46. For carbon dioxide ($CO_2$), $M_r$ = (1 x 12) + (2 x 16) = 44. Now, using the equation given on page 12 and entering the relative molecular mass for each molecule we can calculate:

$$C_6 H_{12} O_6 \longrightarrow 2\ CH_3 CH_2 OH\ +\ 2\ CO_2$$
$$(180) \qquad (2 \times 46 = 92)\ +\ (2 \times 44 = 88)$$

This means that 180 grams of glucose would yield 92 grams of ethanol and 88 grams of carbon dioxide. By ratio, 100 grams of glucose would yield 100 x (92 / 180) grams of ethanol and, therefore, 1000 grams (1 kilogram) of glucose would yield 1000 x (92 / 180) grams of ethanol = 511.1 grams. Thus, the theoretical yield of ethanol from fermentation of 1 kilogram of pure glucose would be 511.1 grams. This is a theoretical yield because 100% fermentation efficiency is unobtainable in practice. Following fermentation, ethanol (as aqueous liquor) is separated from the yeast by filtration. Commercial 95% ethanol is obtained by fractional distillation. For economic reason, sucrose is used for fermentation.

# THE CHEMISTRY OF COOKING

Everyone who cooks is a chemist, with the kitchen as their laboratory, because cooking *is* chemistry. When we cook we use heat to produce chemical reactions between food molecules, changing them into other molecules, which in turn give the cooked food new properties, changing its flavor, smell, color, consistency, nutritiousness, and even toxicity.

When food is cooked, changes in its chemistry occur as a result of heating. Heat causes many of the complex molecules in food to break down into smaller molecules. Some of these changes are important for human consumption of food in that they allow us to digest the food chemicals more successfully—for instance, tough proteins in meat are broken down during cooking into more digestible forms. Heating also facilitates the occurrence of reactions between food chemicals, such as the Maillard reaction and caramelization.

## The Maillard Reaction

One of the most important reactions in cooking was discovered by the French chemist Louis Camille Maillard (1878–1936), who in 1912 found that if enough heat is applied, two of the vital components of food—proteins and carbohydrates—will react together to produce a distinctive new set of molecules. The new molecules give some cooked foods their flavors, scents, and colors.

Another reaction between food molecules that only happens when heat is applied is caramelization, a dehydration reaction where water molecules are driven off from sugars, converting them into new forms of sugar and eventually into other molecules that produce the sensation of flavor, smell, and color. The Maillard reaction and caramelization work together to produce the distinctive flavors, smells and colors of cooked foodstuffs like freshly toasted bread, browned meat, roasted coffee, and popped corn.

Why do humans find these aromas, tastes, and sights so appetizing? Once again, the key is chemistry. Among the chemicals released during cooking are many scent molecules similar to those given off by ripening fruit, a source of energy-rich sugars that would have appealed to our ape-like ancestors as a vital component of their diet. Food scientist Harold McGee suggests that, "fruits probably provided our evolutionary ancestors with refreshing sensory interludes in an otherwise bland and dull diet . . . perhaps cooking with fire was valued in part because it transformed blandness into fruitlike richness."

## Yeast of Eden

Not all cooking involves heat. The kneading of dough in the initial stages of bread-making is an example of mechanical alteration of food chemistry. Kneading causes proteins present in the dough, such as gluten, to align and cross link to form long, elastic chains, which in turn allow the dough to rise by trapping the gas produced by the action of yeast. Yeast is a fungal microorganism that converts sugars into ethanol, a type of alcohol, giving off carbon dioxide in the process, in a reaction known as fermentation. In baking the alcohol is a side-product, and is mostly driven off during baking, but fermentation as an end in itself is the basis of brewing.

Yeast's remarkable chemical abilities were probably first harnessed by prehistoric man, but the first records of

### THE COOKING APE

Richard Wrangham of Harvard University argues that mastery of the chemistry of cooking was the driving force behind human evolution. Cooking makes food easier and quicker to eat and digest, releasing more calories and allowing a wider range of foodstuffs to be added to the diet. Cooked food—and particularly cooked meat—requires less chewing and uses less energy to digest, and Wrangham says that being able to cook allowed early humans to evolve energy-hungry large brains while freeing up time to develop culture, society, and technology. Cooking explains why our ancestor species, *Homo erectus*, evolved a smaller jaw, shorter intestine (and therefore smaller belly), and larger skull (to accommodate that big brain). "Cooking is what makes the human diet 'human,' and the most logical explanation for the advances in brain and body size over our ape ancestors," Wrangham says. "It's hard to imagine the leap to *Homo erectus* without cooking's nutritional benefits."

brewing come from ancient Mesopotamia, ca. 4000 BCE, while beer was also brewed in pre-Dynastic Egypt around the same time. Beer was produced in large enough quantities for export, marking brewing as one of the earliest forms of industrial chemistry.

# MUMMIES, MEDICINES, AND MAKEUP

The word "chemistry" traces its roots back to ancient Egypt, where the civilization that developed in the Nile Valley from 3100 BCE matched the sophistication of its architecture and art with equally sophisticated chemistry. Ancient Egyptians used a wide range of chemicals and knew how to refine and combine them to best effect.

The ancient Egyptian world was one of vivid colors, produced via a mastery of pigments and dyes used to enrich paintings, fabrics, makeup, and glass. To prehistoric staples like ocher and other oxides of iron, the Egyptians added pigments based on cobalt, lead and copper, recruiting more of the elements than had previously been used. Lead, for instance, was mined in the form of galena, an ore composed of lead sulfide (PbS), obtained from Gebel Rasas ("the Mountain of Lead"), a few miles from the Red Sea coast. Mercury, aka quicksilver, was also mined. Egyptian sages began to develop a complex system of lore and mysticism relating to the known metals, linking gold with the Sun, iron with Mars, copper with Venus, and lead with Saturn. Although pre-scientific, this system of knowledge followed its own rules and was internally coherent and even rational, and so might be said to mark the beginnings of chemistry.

## ORIGINS OF THE WORD "CHEMISTRY"

The word "chemistry" comes from alchemy (see pp. 36–37), the Western pronunciation of the Arabic al-kimya, but the origins of the root word kimya vary according to the source. The Roman natural philosopher, Pliny the Elder, claimed it derived from the ancient Egyptian kemi, meaning "black," as in the black silt of the Nile, the first matter in the Egyptian cosmology, and the name given to Egypt itself, and thus by extension to the "Egyptian art." Alternatively it was said to derive from the Greek word khemeia, meaning "pour together" (referring to the fusion of molten metals).

## Egyptian Blue and Royal Purple

Silicon is the second most abundant element in the Earth's crust after oxygen, and the Egyptians put it to good use; as early as the eighteenth dynasty (16th century BCE) they had developed furnaces hot enough to melt it and produce glass, and they later learned how to add lead to glass to increase its refractive index (making it sparkle).

A technology parallel to glass was the manufacture of faience, a ceramic made from a paste of crushed quartz (an oxide of silicon in crystal form) or sand. To this was added small amounts of lime (calcium carbonate) and natron, a naturally occurring mixture of sodium carbonate, sodium bicarbonate, and common salt, which functioned as a sort of secret ingredient in Egyptian chemistry. Natron acted as a flux, supplying positively charged ions that helped break up the rigid lattice in which the silicon atoms were arranged, lowering its melting point. Faience was glazed with copper pigments to give a bright blue-green color, forming an artificial substitute for rare and expensive lapis lazuli.

Natron and silicon were also used by the Egyptians to create an entirely new color, Egyptian blue. This artificial pigment was made by heating a mixture of sand, natron, and copper filings to about 1500°F (850°C). At this temperature the raw materials would melt and react to give a new compound, copper calcium tetrasilicate. From the Levant the Egyptians obtained royal purple, a reddish-purple dye made from a snail.

Other pigments known to have been used in Egypt at least as early as 2650 BCE included copper carbonate, limestone, and charcoal.

## Kill and Cure

Pharmaceutical use of chemicals probably pre-dates humanity, as there is evidence that apes self-dose with pharmacologically active plants, but the chemistry of medicine reached new heights under the Egyptians, who made use of many chemicals that remained in the pharmacopoeia until the 19th century. The Ebers papyrus, one of the two oldest medical treatises anywhere (it dates from 1534 BCE, but is based on much older sources), shows that the ancient Egyptians were familiar with medicines based on lead and antimony (such as stibnite—black antimony sulfide—for fevers and skin conditions), as well as a host of plant extracts such as opium and aconite.

When remedies failed, the Egyptians were experts in the chemistry of the mortuary arts. Their recipe for mummification made use of the properties of natron, which absorbs water to dehydrate the corpse and also has antibiotic properties. Sodium carbonates are alkalis (see pp. 70–71), raising the pH of the treated flesh, which also helps retard bacterial growth. The dried corpses were then treated with pitch and tars, such as bitumen, to seal and preserve still further. In the right conditions mummies can survive intact for at least 3,000 years.

# 2 Determination of Lead

## THE PROBLEM:

It is known that Queen Nefertiti and other ancient Egyptian royals used lead-based substances in the formulation of cosmetics, particularly as an ingredient in black eye makeup (kohl). However, nowadays, lead compounds are banned in cosmetics on the grounds of their high toxicity and so analysis for lead in imported cosmetics is very important. How can we determine the percentage amount of lead in ancient Egyptian cosmetics?

## THE METHOD:

Recent chemical analysis of cosmetics found in ancient Egyptian tombs, and the reconstitution of ancient recipes, have shown that two non-naturally occurring lead chlorides, laurionite (Pb [OH] Cl) and phosgenite ($Pb_2 Cl_2 CO_3$), must have been synthesized by the ancient Egyptians and used as fine powders in makeup and eye lotions. According to ancient Egyptian manuscripts, these were essential remedies for treating eye illness and skin ailments. As in Exercise #1, we can use the atomic weight ($A_r$) for carbon (C = 12), hydrogen (H = 1), oxygen (O = 16), chlorine (Cl = 35.5) and lead (Pb = 207) to calculate the relative molecular masses ($M_r$) for laurionite and phosgenite. From this we can calculate the percentage amount of lead (Pb) in both substances.

Laurionite (Pb [OH] Cl) contains 1 x Pb, 1 x H, 1 x O and 1 x Cl atoms. Therefore, the relative molecular mass of laurionite is:

$$(1 \times 207) + (1 \times 1) + (1 \times 16) + (1 \times 35.5)$$
$$\text{Pb} \qquad \text{H} \qquad \text{O} \qquad \text{Cl}$$

$$= 259.5$$

Phosgenite ($Pb_2Cl_2CO_3$) contains 2 x Pb, 2 x Cl, 1 x C and 3 x O atoms; therefore, the molecular weight is:

$$(2 \times 207) + (2 \times 35.5) + (1 \times 12) + (3 \times 16)$$

Pb          Cl          C          O

$$= 545$$

## THE SOLUTION:

To find the percentage of lead in laurionite, we simply divide the atomic weight of the lead atom by the molecular mass of laurionite:

$$(207 / 259.5) \times 100 = 79.77\% \text{ Pb}$$

By the same process, we can calculate the percentage of lead in phosgenite:

$$[(2 \times 207) / 545] \times 100$$

$$(414 / 545) \times 100 = 75.96\% \text{ Pb}$$

This means that laurionite and phosgenite contain 79.77% Pb and 75.96% Pb respectively. Ingested lead compounds are highly toxic because they react with stomach acids to form soluble $Pb^{4+}$ ions, making these Egyptian cosmetics unsuitable for the modern market!

# MATTER AND ENERGY

Before we go any further, let's take a quick look at some basic concepts and terminology. Chemistry is the science of matter—its makeup and transformations—so its fundamental concepts concern the description of matter. Chemical transformations involve energy, so concepts relating to energy form the other conceptual pillar of chemistry.

## Phases of Matter

Matter is that part of the universe that has mass and occupies space; it is the material of which the physical world is constructed, the stuff you can see and touch. Matter can exist in three states, known as phases: solid, liquid, and gas.

A solid has a definite shape and volume, because the particles that make it up, whether atoms or molecules, are held close together in a relatively rigid, immobile structure by powerful bonds (which may be covalent or ionic; see pp. 78–79). Sometimes this structure involves repeating arrays of the exact same pattern, known as a crystal lattice. Crystalline solids include ice, table salt, granulated sugar, and quartz. The particles in a solid are not completely motionless; they vibrate in place, but remain fixed relative to one another.

When a solid is heated to its melting point it becomes a liquid, the phase in which matter has no definite shape but does still have a definite volume—in other words the liquid remains as a definite body or mass. In a liquid the bonds, or forces of attraction, between particles are stronger than in a gas but much weaker than in a solid, allowing the particles to move about.

When a liquid is heated to its boiling point it undergoes a further change of state, becoming a gas, the phase in which matter has no definite form or volume. In a gas the forces attracting particles together are so weak that the particles can move about freely. As a result a gas will expand to fill the volume available.

Melting and boiling are the terms used to describe phase changes from solid to liquid and liquid to gas. The reverse phase changes are known as freezing (going from a liquid to a solid) and condensation (going from a gas to a liquid). Some substances go directly from solid phase to gas phase; this is known as sublimation. Solid (frozen) carbon dioxide, known as dry ice, is an example of a

solid state

liquid state

gas state

Turn to pp. 98–99 for more on the relationship between heat, temperature, and the phases of matter.

substance that sublimates. The smoky vapor given off by a block of dry ice, however, is not gaseous carbon dioxide, which is colorless, but water vapor condensing out of the air as it is cooled by the subliming carbon dioxide. Phase changes are caused by the amount of energy the particles have. High levels of energy enable particles to break free of the bonds that hold them together in their solid phase and transition to liquid or gaseous phases. As a substance is cooled the particles lose energy and the attractive forces between the particles reassert themselves.

## Properties of Matter

Matter can be pure or mixed. In a mixture, different substances are physically combined—for instance, if you mix together chalk dust and salt, or dissolve salt in water. A pure substance has a single, constant composition, and can be either an element or a compound (see p. 6). Pure substances have chemical and physical properties, and these are what chemists study. Chemical properties include how reactive a substance is, what it will react with, and other factors governing the substance's transformation into other substances. Physical properties include mass, dimensions, volume, density, color, conductivity, and so on. Physical

## ENERGY

Along with matter, energy is one of the two basic components of the universe. Energy can take different forms; the most important forms in chemistry are kinetic and potential energy. Kinetic energy is the energy of motion that particles have, and determines the speed and force of their motion, governing properties such as phase and reactivity. Potential energy is the energy stored in a substance, which can become other types of energy. In chemistry the type of potential energy of most interest is the energy stored in the form of chemical bonds. Breaking the bonds takes energy but can also release energy. Except in nuclear reactions, energy cannot be created or destroyed, only converted from one form into another.

properties are described with standard units of measurement, such as the gram, meter, and liter. Smaller units are created by adding prefixes, such as "centi-" (meaning a hundredth of) and "milli-" (thousandth of). A milligram is therefore 0.001 grams. Volume is measured in cubic meters (or centimeters, or millimeters)—also known as $m^3$ (or $cm^3$ or $mm^3$)—or liters, L, (more commonly milliliters, ml). Density is mass divided by volume, and is therefore typically measured in g/ml.

# NATURAL PHILOSOPHY IN ANCIENT GREECE

Around the 5th century BCE, new ways of thinking about the natural world emerged in the ancient Greek world. Although older civilizations such as the Egyptians and Babylonians had extensive practical experience of chemistry from medicine to metallurgy, they had made no attempt to investigate systematically the natural world. This was about to change with the development of the first theories concerning the makeup of matter.

## Evolving Elements

The ancient world made use of gold, silver, tin, lead, copper, iron, mercury, antimony, sodium, calcium, carbon, sulfur, and arsenic in one form or another, yet they were not recognized as elements and not until the Greeks was there any attempt to analyze and explain the differences between these and other forms of matter. Among the first to use evidence from the natural world to pose and answer fundamental questions about the nature of matter was the natural philosopher Thales of Miletus (ca. 625–ca. 547 BCE), a semi-legendary figure from a Greek city-state in what is

### • THE BAGHDAD BATTERY

An ancient jar dug up near Baghdad in 1939, dated to around 200 BCE, proved to contain a copper cylinder, within which was an iron rod, both of them held in place in the neck of the jar by an asphalt plug. Almost immediately it was suggested that it might be a form of battery used for electro-plating (where a thin layer of one metal—gold or silver, for example—is deposited onto the surface of another metal, in order to gild or silver it). The "Baghdad Battery" is now among the best known pieces of anachronistic technology, apparent evidence that the ancients had sophisticated electro-chemical knowledge. In practice, however, the Battery and its sisters (more than a dozen have been unearthed) may not be what they seem, lacking design elements necessary for a working battery. They are now believed to have been containers for scrolls of papyrus or bronze, dating to the Sassanian era (225–640 CE).

now Turkey. Thales is credited as being the first man to seek naturalistic explanations instead of attributing phenomena to the gods. In particular he identified water as the first principle, the basic element from which all other matter is composed.

Thales was followed by his pupil Anaximenes (whose career flourished ca. 546–526 BCE), who argued that the fundamental element was in fact air, and that through processes of condensation and rarefaction it became earth, fire, water, and all the other forms of matter. Reacting against this theory, Heraclitus of Ephesus (ca. 500–475 BCE) proposed that nature was flux, with constant change the true order of things; accordingly the basic element must be immaterial and mutable, and he identified fire as the first principle.

Finally, Empedocles (ca. 492–432 BCE) claimed that there were four basic forms of matter—earth, air, fire, and water. Thanks to its adoption by Aristotle (see pp. 28–29), this "four element" model would become the accepted standard for over 2,000 years. It can be argued that the four elements correspond to the modern model of the universe, with three phases of matter (earth = solid; water = liquid; air = gas), together with energy (= fire).

## HERO'S AEOLIPILE

Hero of Alexandria (62–152 CE) was a Greek natural philosopher and inventor who was millennia ahead of his time in the field of pneumatic chemistry—the study of gases (see pp. 64–65). He demonstrated that air was a material substance, experimented with compression, and theorized on vacuums. He is most famous for his aeolipile, a device he built using his knowledge of the phase change between water and steam: a cauldron of boiling water supplied steam to a sphere with two vents at right angles. As the steam was forced out under pressure so the sphere rotated. It was the first steam engine, a technology based on pneumatic chemistry that would later reshape the world by triggering the Industrial Revolution. One of the great questions of the history of science is why Hero's invention failed to trigger a comparable revolution in his own time. The widespread use of slaves is usually suggested as the answer— there was no need for labor-saving devices at this point in history.

• An illustration of the aeolipile: the fire heats the water in the cauldron, supplying steam to the sphere via the pipes, which also act as an axle.

# THE BIRTH OF ATOMIC THEORY

The foundation of modern chemistry is atomic theory—the model that explains what substances are made of and how and why they come together to form molecules. Atomic theory in its modern form dates to the 19th century but drew on an ancient but largely forgotten tradition—the Atomists of ancient Greece.

## Atoms and Empty Space

A school of Greek philosophers known as the Eleatics argued that since nothingness is a logical impossibility, there cannot be any space between particles and thus no discrete, indivisible particles. This complex argument led to apparently irrational conclusions; for instance, the Eleatics claimed that change is impossible and nothing could come into or go out of being, and even that real motion itself is impossible. Reacting against this school of thought, the philosopher Leucippus (5th century BCE) and his pupil Democritus (ca. 460–370 BCE) claimed that empty space (what we would today call a vacuum) can exist, and therefore it is possible for particles to exist.

These particles of "being" fulfill the criteria of the Eleatics: they are unchangeable and indivisible, known as atoms (from the Greek *atomos* meaning "uncuttables"). In other words, if you cut a piece of matter into smaller pieces, and then cut up those pieces, and so on, you will eventually reach the smallest unit possible—a particle that cannot be cut into smaller particles. According to this "atomic theory," atoms are solid and so small that they are invisible, but they come in different shapes and sizes and can change position. Different arrangements and combinations of atoms produce different materials, even different worlds, according to Democritus.

"indivisible"

• Democritus, proponent of the atomic theory, known to later generations as "the laughing philosopher."

"The first principles of the universe are atoms and empty space. Everything else is merely thought to exist. The worlds are unlimited . . . [atoms] generate all composite things—fire, water, air, earth."—*Democritus*

## Atomism Eclipsed

The Atomists, as Leucippus, Democritus, and their later supporters were known, appear to have been startlingly prescient, anticipating modern thinking about atoms, elements, and cosmology. However, we should not get too carried away; their model was just speculation, and not based on a true scientific approach (see pp. 24–25). It included mystical—or metaphysical—ideas, such as the belief that the human soul was also made of atoms (particularly fine, round ones).

Although Atomism had its supporters in ancient Greece, it was rejected by the most influential of later philosophers, such as Plato and Aristotle, and did not become popular again until the Scientific Revolution of the 17th and 18th centuries. Whether the insights of Democritus into the nature of matter would have speeded the advent and advance of scientific chemistry is impossible to say, but the theories that superseded Atomism, particularly the ideas of Aristotle, have been held responsible for leading chemistry up a 2,000-year-long blind alley.

### THE PHILOSOPHER WHO FELL OFF A HILL AND OTHER STRANGE DEATHS

Although his theories failed to attract the support they perhaps deserved, leading him to complain that when he came to Athens "no one knew me," Democritus himself was fortunate to live to a great age. Possibly this was thanks to his personal philosophy, which was that cheerfulness should be the purpose of life; to later generations he was known as "the laughing philosopher." Many of the other ancient Greek philosophers mentioned here were not so lucky. Thales of Miletus was said to have fallen off a hill because he was so intent on studying the stars. His pupil Anaximenes may well have been killed by invading Persians. Heraclitus met perhaps the most colorful end— because of his philosophical principles he kept to a starvation diet; this in turn caused him to swell up with dropsy, and seeking to draw out his "ill humors" he buried himself in a dung heap and never came out.

# INTRODUCING ATOMS

At the scale of matter investigated by chemistry, the basic building block is the atom. Atomic structure determines the properties and chemistry of a substance. This structure is comprised of subatomic particles, for it is now known that Democritus and the ancient Atomists were wrong, and that far from being indivisible, atoms actually have complex internal workings.

## Atoms and Elements

An atom is the smallest particle that still represents a particular element. For instance, the smallest particle of gold that is still gold is an atom of gold. If an atom of gold is broken into its constituent parts it will no longer be gold. Any atom of gold is identical to any other atom of gold (excepting different isotopes—see pp. 62–63). Each element has a unique atomic structure; it is this structure that defines the element and determines its nature.

## Subatomic Particles

Atoms are composed of three types of subatomic particle: the proton, the neutron and the electron. The proton and the neutron are far larger than the electron. Over 99.99% of an atom's mass is made up of protons and neutrons. A proton has 1836 times the mass of an electron. The number of protons an atom has determines its atomic number, while the combined total of protons and neutrons determines its mass number (see pp. 120–121 for more on atomic number and mass).

These subatomic particles may be electrically charged. Protons are positively charged (with a charge of +1), while electrons are negatively charged (with a charge of −1). Neutrons, as their name implies, are neutral, with no charge (a charge of 0). An atom is electrically neutral, because the number of electrons it has matches the number of protons. For instance, a helium atom has two protons and two electrons, while a uranium atom has 92 protons and 92 electrons. If an atom loses or gains an electron, so that it has more or fewer protons than electrons, it may become positively or negatively charged, in which case it is known as an ion.

## The Orbital Model

The simplest useful model of the internal structure of the atom is the orbital model developed by Danish scientist Niels Bohr. In this model the atom is like a tiny solar system. At the centre is the nucleus, where the protons and neutrons are packed together in a dense mass; indeed this is where the vast majority of the atom's mass is located.

Orbiting around the nucleus like tiny planets are the electrons. If there are more than two electrons they do not all orbit at the same distance. The various orbits—or shells—represent different energy levels, with the shell nearest the nucleus having the lowest energy level. Electrons can move up and down between shells, but there is only so much space in each shell. This arrangement determines the valency of the atom, which in turn determines many of its chemical properties (see pp. 78–79 for more on these concepts).

In practice the orbital model is an oversimplification. Closer to what scientists see in real life is the quantum mechanical model, in which, thanks to the Uncertainty Principle, it is impossible to know both the position and momentum of an electron at the same time. Accordingly electrons are said to occupy fields of space known as orbitals or electron clouds.

• Diagrammatic view of the electron configuration of gold (Au). With as many protons as electrons, gold atoms are extremely heavy and the metal itself is extremely dense.

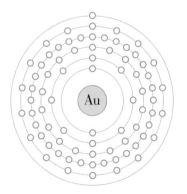

## Two's Company

Although an atom is the smallest unit of an element, for most of the history of chemistry it was not possible to decompose every element to single atoms. Some elements, even in their pure state, when not compounded or mixed with any other atoms, favor binding to another atom. For instance, oxygen is never encountered as single atoms of oxygen—even in a jar filled with nothing but pure oxygen, the element will exist as pairs of atoms, bound together to give a diatomic molecule. Hydrogen behaves in a similar diatomic form, as do five other elements: nitrogen, fluorine, chlorine, bromine, and iodine. This phenomenon caused many headaches for 18th- and 19th-century chemists attempting to determine atomic numbers and masses.

# Aristotle

Towering over the history of science from ancient times to the early modern period is the colossus Aristotle, whose model of the physical world and theory of the elements were accepted as standard until the 16th century and beyond. Student of Plato and tutor to Alexander the Great, Aristotle became a legendary figure, yet for chemistry his legacy was double-edged at best.

## Students and Masters

Aristotle (384–322 BCE) was born in Macedonia, where his father was a physician at the royal court. At the age of 17 he traveled to Athens to study with Plato at the Academy. He must have been a brilliant student, and some historians trace developments in Plato's thinking to criticisms made in Aristotle's own writings, suggesting that the older man took notice of points made by his pupil. Aristotle remained at the Academy until his mentor's death in 347 BCE, after which he took up positions at Assos in Asia Minor and Mytilene on Lesbos, during which time he pursued research into marine biology that was unmatched until the modern era (Darwin was a great fan). In 342 he was headhunted by Philip of Macedon to act as tutor to the young prince, Alexander.

In 335 Alexander came to the throne and Aristotle returned to Athens to found a school of his own, the Lyceum. But anti-Macedonian feeling ran high in the city and when Alexander died in 323 Aristotle feared that he would end up like Socrates (the great philosopher had been forced to drink poison after being falling foul of public opinion). Declaring that he did not want Athens to "sin twice against philosophy" he moved to the city of Chalcis where he died the next year.

• **ARISTOTLE'S LEGACY**

Aristotle's theories about the elements were so important because of the immense influence his work would exert on natural philosophy, particularly in Europe, over the next 1900 years. For a variety of reasons his system of logic and physics appealed to the Church, and this in turn enshrined his popularity and authority. By the Middle Ages Aristotle was the focus of Scholasticism, a system of thinking that dominated the intellectual life of Europe, and he was taken as the last word on natural philosophy, including matters chemical. "The reading of Aristotle will not only conduce much to your study . . . but also help you in Greeke, & indeed crown all your other learning," wrote Richard Holdsworth in his "Directions for a Student in the University," a 17th-century study guide.

## The Fifth Element

Plato had largely rejected the Atomist view of matter, teaching that physics was dependent on metaphysics, in as far as the forms and substances of the material world were inferior copies of ideal forms that existed on a higher plane. This had little practical utility for natural philosophy, reflecting Plato's distaste for the realities of the material world—and for the practice of actually observing the world, rather than simply thinking about it. One of Aristotle's most important innovations was his willingness to get his hands dirty in some fields of research (such as his investigation of animal biology).

Yet the core of Aristotle's philosophical project was his system of logic, in which knowledge was advanced through logic and rational thought alone. He formalized a system of deductive logic, which proceeded by syllogisms. In a syllogism you start with premises and use them to deduce conclusions. But if the premises are faulty, then the conclusions will also be faulty. This may explain how Aristotle came to believe that the universe was made up of five elements.

Accepting the four terrestrial elements of Empedocles, he added a fifth one—ether—to explain the workings of the heavens. For Aristotle the natural qualities of the elements explained everything about the material world. Earth was naturally heavier than air, so substances with a higher proportion of the earth element would naturally fall until they were lower than airy substances. Substances with a high proportion of the fire or water element would have "hot" or "damp" qualities, respectively, and this explained their chemistry. Aristotle had mistaken qualities for properties (in the scientific sense), and this fundamental error informed his premises and thus rendered his conclusions incorrect.

He made similar errors in other fields, led astray by failure to back up logic with observation. For instance, relying on pure deduction he concluded that the function of the brain is to cool the blood and that man has only eight ribs on each side. Today it seems incredible that Aristotle could have made simple mistakes, such as his claim that women have fewer teeth than men; as the 20th-century philosopher Bertrand Russell observed, "all he had to do was to ask Mrs. Aristotle to open her mouth and count them."

### ARISTOTLE AND ALEXANDER

Not much is known about the relationship between Alexander and Aristotle. It is recorded that Alexander took with him on his campaigns a copy of Homer's *Iliad* with annotations by Aristotle, while Plutarch quotes a letter from Alexander rebuking his teacher for publishing material the great king had believed was for his ears only. Aristotle was said to have taught his royal charge politics and ethics; when he later wrote on the subject of kingship, suggesting that a man was only fit to be absolute ruler if he was effectively superhuman, he neglected to mention Alexander's name as an example of such a paragon.

# 3 Estimation of Nitrogen in Gunpowder

## THE PROBLEM:

The power of gunpowder was discovered by accident in the 9th century by Chinese alchemists who hit on an explosive recipe of 75 parts of saltpeter (potassium nitrate) with 15 parts of charcoal (carbon) and 10 parts of sulfur. Gunpowder is explosive because of the rapid generation of heat and expansion of trapped carbon dioxide ($CO_2$) and nitrogen ($N_2$) gases liberated in its reaction. The presence of nitrogen in saltpeter plays an important part in making gunpowder so explosive; how can we find the amount of nitrogen (as a percentage) in a 1 kilogram sample of gunpowder?

## THE METHOD:

Over time, the composition of gunpowder has varied and sodium nitrate (Chile saltpeter) is sometimes used in place of potassium nitrate, particularly in fireworks. In this exercise, however, we will calculate the nitrogen content of two gunpowder mixtures that use potassium nitrate:

(a) 75 parts of saltpeter ($KNO_3$) mixed with 15 parts of charcoal (C) and 10 parts of sulfur (S)

(b) 67 parts of saltpeter mixed with 22 parts of charcoal and 11 parts of sulfur

In the equation for this explosive gunpowder reaction, potassium nitrate reacts with carbon and sulfur to produce nitrogen, carbon dioxide and potassium sulfide ($K_2S$):

$$2\ KNO_3(s) + 3\ C(s) + S(s) \rightarrow$$
$$N_2(g) + 3\ CO_2(g) + K_2S(s)$$

Note that (s) indicates a solid state and (g) a gas.

As we are only interested in the amount of nitrogen in the two mixtures, the carbon and sulfur elements in the equation can be disregarded. We need only the atomic weights for potassium ($K = 39$), nitrogen ($N = 14$), and oxygen ($O = 16$) to find the relative molecular masses for potassium nitrate. From this we can calculate the percentage of nitrogen in 1 kilogram of each of the two mixtures.

## THE SOLUTION:

The chemical formula of potassium nitrate is $KNO_3$. Its relative molecular mass is:

$$(1 \times 39) + (1 \times 14) + (3 \times 16) = 101$$
$$\quad K \qquad\qquad N \qquad\qquad O$$

In 1 kilogram of mixture (a), 75 parts of saltpeter equates to 750 grams. Again, we don't need to know the amount of carbon or sulfur. To start, we must calculate the percentage of nitrogen in a molecule of potassium nitrate, using their respective atomic and molecular masses:

$$(14 / 101) \times 100 = 13.87\%$$

Now we need to place this in context of a 1 kilogram mixture that contains 750 grams of potassium nitrate:

$$(750 / 1000) \times 13.87 = 10.40\%$$

Therefore, 1 kilogram of gunpowder mixture a contains 10.4% nitrogen.

To calculate the amount of nitrogen in mixture (b), we simply have to adjust the previous calculation to reflect the different amount of potassium nitrate in the mixture:

67 parts in a 1 kilogram mixture equates to 670 grams. The calculation is adjusted thus:

$$(14 / 101) \times 100 = 13.87\%$$

$$(670 / 1000) \times 13.87 = 9.29\%\ N$$

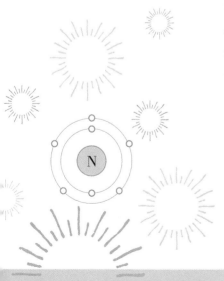

# GREEK FIRE OF THE BYZANTINES

A remarkable instance of ancient chemical technology was the secret weapon that preserved the Byzantine Empire for 600 years, the mysterious "Greek fire." The recipe for this extraordinary chemical weapon was among the most closely guarded secrets in history, and with good reason, for it had the power to change the very course of history.

## Smells Like Victory

Greek fire was a napalm-like incendiary substance used in the defense of the Byzantine Empire. Its technology was a closely guarded secret known only to the imperial family and associates, and remains an enigma to this day. It was first used in 678 CE against the Arabs, who were threatening to overrun Constantinople, having already conquered the Persians. Although the defenses of the capital, primarily the colossal Theodosian Walls, could hold back land armies, the city could be starved into submission if the Arab fleet could gain control of the seas.

But the Arabs had sewn the seeds of their own defeat. When their armies overran Christian Syria, refugees flocked to the safety of Constantinople. Among them was a Syrian Greek called Kallinikos, who brought with him the recipe for a secret weapon that would become known as "Greek fire," in reference to his ethnicity, although it has also been known as "liquid fire," "sea fire," or "Persian fire." The last of these is a possible reference to its true origin, for sources claim that Kallinikos had previously been in the employ of the Muslim military. Incendiary weapons based on petroleum products, such as pitch or naptha, were part of the Arab arsenal; in fact they were probably known in one form or another to the Romans and Persians before them. What distinguished the new "Greek" fire was its advanced composition, and, crucially, the "delivery" technology—the apparatus used to spray the flaming liquid toward the enemy.

• Illustration from a Byzantine manual on siege warfare, showing a handheld version of the siphon used for dispensing Greek fire.

## DISCLOSED BY AN ANGEL

Greek fire was used as little as possible, to help prevent the siphon apparatus falling into enemy hands, while prohibitions and legends grew up around its mysterious recipe. Writing to his son, Emperor Constantine VII Porphyrogennitos stressed that the secret must not be revealed even to allies, explaining: "The ingredients were disclosed by an angel to the first great Christian emperor, Constantine ... [who] ordained that they should curse, in writing and on the Holy Altar of the Church of God, any who should dare to give this fire to another nation ... let him be deposed and paraded like a common criminal throughout the centuries, whether he be an emperor, a patriarch, or any other lord or subject."

## Liquid Death

Even today it is only possible to speculate on the composition of Greek fire, but it is generally thought to have included sulfur, quicklime, liquid petroleum, and perhaps even magnesium (a constituent of modern incendiary weapons). Magnesium is a highly reactive metal that will even burn underwater, one of the characteristics attributed to Greek fire, which helped to make it such a fearsome weapon. To spray this liquid death, the Byzantines invented an ingenious siphon device.

The effects of Greek fire were devastating. In 678 the deployment of Greek fire dramatically turned the tide of battle against the Arabs; their navy was shattered, with the loss of thousands of men. The siege was broken and the Arabs were forced to sue for peace. When they attacked again, in 717, Greek fire once again played a pivotal role in the defense of the city, and the Arabs were again beaten back with severe losses.

Over the next three centuries Greek fire was invaluable in defence of the Byzantine Empire, but by 1204 the secret had somehow been lost, the chain of transmission of secret knowledge broken. Incendiary weapons were still used (and still referred to as "Greek fire"), but the package of technology that made Greek fire so formidable was no longer available. The empire fought on for another five centuries, until in 1453 the Ottoman Turks breached the walls of Constantinople using gunpowder.

But Greek fire had already made its impact on history, checking the hitherto unstoppable spread of Muslim armies and holding back the forces of Islam in the Eastern Mediterranean for hundreds of years. If the Arabs had overrun the Byzantine Empire in the 7th century, how far might they have penetrated? The shape of Europe and the direction of world history might have turned out very differently without the secret of Greek fire.

# Alchemy and the Birth of Chemical Science

The chemical legacy of the ancient world was a
colorful stew of materials and mysticism that proved
irresistible to scholars through the ages, drawing the
greatest minds of late Antiquity, medieval Islam and
the European Renaissance like moths to a flame.
Their efforts to unlock the secrets of the cosmos would
lead to a new way of investigating nature and toward a
new science of matter and its transformations.

# THE ROOTS OF ALCHEMY

The city founded in Egypt in 331 BCE by Alexander the Great quickly became representative of his new world, a thriving, cosmopolitan blend of races, cultures, and traditions. It was in Alexandria that chemistry would begin to emerge in the form of alchemy—an art rather than a science, but recognizable in many of its practices and concerns.
Like the city itself, this new art was a complex hybrid.

## The Egyptian Art

In Alexandria under the Ptolemaic dynasty, Greek (or rather, Hellenic) culture was grafted onto ancient Egyptian traditions. Magic, mysticism, and the chemical arts that the Egyptians had practiced for millennia, such as embalming, glass-making, faience, and metallurgy, went into the crucible with Hellenic metaphysics and Aristotelian cosmology. What emerged was a strange new compound.

Alchemy adopted the Classical elements (earth, air, fire, water, and ether), and sought to make use of Aristotelian theories about matter. If the nature of a substance was determined by its particular proportions of elements, as Aristotle said, altering those proportions would change the nature of the substance. If gold, for instance, was formed by one particular recipe of earth, air, fire, and water, then taking a baser metal such as lead, and altering its recipe until it matched that of gold, should effect the

## • HERMES TRISMEGISTUS

*The legendary, semi-divine figure to whom alchemy was attributed was Hermes Trismegistus ("thrice-great"), himself a composite of the Egyptian god of wisdom, Thoth, and the Greek god Hermes. Hermes was said to have penned a body of work known as the* Hermetic Corpus *in the dim recesses of Antiquity; all subsequent alchemical research and learning was merely an attempt to decipher and recover this original wisdom.*

transmutation of lead into gold. Seeking the means to perform this transmutation, alchemists worked with substances known from Egyptian, Greek, and Roman technology: the metals, earths such as ocher, and ores such as stibnite, performing operations such as solution, distillation, and filtration.

Guiding these operations were mystical laws, especially "as above, so below." This saying represented a belief that the microcosm—the "little" world of man and earthly matter, including the substances that were the alchemist's stock in trade—corresponds to the macrocosm—the universe at large, including the stars and heavenly bodies. This law gives rise to the Doctrine of Correspondences, the belief that everything in the microcosm, or terrestrial sphere, corresponds to things and phenomena in the macrocosm, or heavenly sphere. The seven known metallic elements, for instance, were held to correspond to the seven "planets" (the generic term for the known heavenly bodies), with gold corresponding to the Sun, silver to the Moon, copper to Venus, and so on. Equally, plants, gemstones, star signs, and all other natural and human phenomena fitted into a network of correspondences, and these could be used to influence the substances that alchemists worked with.

## Why Alchemy Was Not a Science

Such mystical wisdom was believed too powerful and exclusive to be shared with the uninitiated, so alchemists used symbolism and allegory to record their

## ALCHEMY IN CHINA

China had its own alchemical tradition, at least as ancient as the Western one. Chinese alchemists were particularly concerned with the prolongation of life, and a number of legendary Taoist sages were said to have discovered elixirs of immortality. As well as this "internal alchemy" dedicated to human health and wellbeing, the Chinese practiced "external alchemy" with similar aims to the Western version— namely the production of gold. A probable side-product of alchemical research was gunpowder; according to the first recorded account from a Taoist alchemical text ca. 850 CE, it was an explosive discovery: "Some have heated together the saltpeter, sulfur, and carbon of charcoal with honey; smoke and flames result, so that their hands and faces have been burnt, and even the whole house burnt down."

knowledge. Secrecy over methods and results was just one way in which alchemy was unscientific. In accordance with Aristotelian thinking, authority and syllogistic logic took precedence over observation. The practice of alchemy was highly subjective and often mystical in nature—achieving a good result depended on subjective variables such as the experimenter's state of spiritual purity, while substances were affected by the phase of the moon or the position of the stars.

# Gold and Silver

## THE PROBLEM:

Gold (Au) occurs as a native metal, either pure or alloyed (with silver, for example), and also in metal ore deposits. Electrum is a naturally occurring alloy of gold and silver, with trace amounts of copper and other metals, from which gold (and silver) can be processed in a pure form. Mike is planning to propose to his girlfriend. He has bought an 18-carat gold ring and is wondering how much gold this actually contains. How can he determine the amount of gold used?

## THE METHOD:

In order to extract gold, the ore or alloy is mined, crushed and then dissolved in sodium cyanide (NaCN) solution, in the presence of air. The solution is filtered and then the gold is deposited by adding powdered zinc (Zn) metal. This produces a crude gold metal, which is then smelted and finally refined to purity of 99.9999%.

The key to Mike's problem is understanding that the percentage content (by weight) of gold used in jewelry is classed in "carats."

• Gold has been valued as a decorative precious metal for millennia. As the least reactive metal it also has practical value, for instance as an oxidation-proof conductor.

Pure gold (99.9999% Au) is classified as "24 carat"; from this we can find the percentage content of 22-, 21-, 18- and 9-carat gold, and subsequently the gold content in grams of a particular ring.

## THE SOLUTION:

By definition, the carat rating of a material (C) is 24 x Mg / Mm, where Mg is the mass of pure gold in the material and Mm is the total mass of the material. Pure gold (99.9999% Au) is classified as "24 carat."

To calculate 1 carat, therefore:

$$1 / 24 \times 99.9999 = 4.17\%$$

In this way, calculating the gold content of the various grades is straightforward:

22 carat contains (22 / 24) x 99.9999 = 91.7% gold

21 carat contains (21 / 24) x 99.9999 = 87.5% gold

18 carat contains (18 / 24) x 99.9999 = 75.0% gold

9 carat contains (9 / 24) x 99.9999 = 37.5% gold

Pure gold is soft and in jewelry is normally alloyed with other metals to make it harder: the lower the carat rating is, the harder is the piece of gold jewelry.

• In a sort of precursor to the modern system of chemical notation, alchemists used symbols with a rich heritage of occult and astrological meaning to denote important elements.

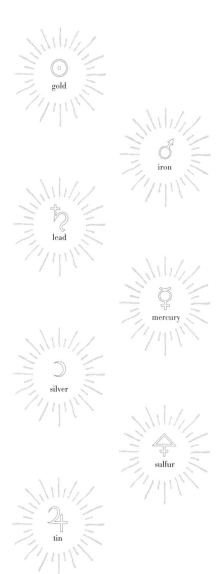

gold

iron

lead

mercury

silver

sulfur

tin

# CHEMICAL REACTIONS 101

In their quest for the transmutation of substances, the alchemists were making a category error. They thought they were achieving what is today known as a nuclear reaction—the transformation of one element into another element. In practice they were performing chemical reactions, the processes by which compounds (elements bonded together in various forms) are created, destroyed, or changed.

## Reaction Basics

In a chemical reaction a substance or mixture of substances is changed into different substances. The substances that exist at the start of the reaction are known as reactants, and the substances that are left at the end are called products. Chemical equations are used to show the reactants and products; they have a before-and-after form like this:

reactants $\longrightarrow$ products

The arrow in the center shows the direction in which the reaction proceeds. Symbols familiar from arithmetic, such as + signs, are used as symbolic shorthand:

reactant A + reactant B $\longrightarrow$ product AB

For example, rusting is an example of a chemical reaction, in which iron combines with oxygen to form iron oxide (rust). The chemical reaction for rust can be written like this:

iron + oxygen $\longrightarrow$ iron oxide

Here is another example, showing what happens when you light the gas on your gas stove:

methane (g) + oxygen (g) $\longrightarrow$ carbon dioxide (g) + water (g)

The letter in parentheses indicates the phase of the substance—in this case all the reactants and products are gases.

Here is an example of a reaction that an alchemist would have performed—the reduction of silver:

silver carbonate (s) $\xrightarrow{\text{heat}}$ silver (l) + carbon dioxide (g) + oxygen (g)

The reaction that gave Chinese alchemists such a nasty shock in the 9th century was:

| sulfur (s) + | carbon dioxide (g) + |
| carbon (s) + $\rightarrow$ | pot. sulfide (s) + |
| pot. nitrate (s) | nitrogen (g) |

Writing out the names of elements and compounds is both laborious and limiting in terms of language (names for substances may change in translation). Accordingly the scientific community adopted innovations such as universal names based on Greek, and scientific notation to improve accuracy and save time (see pp. 114–115).

Reactions that generate heat are known as exothermic, while those that absorb energy are known as endothermic. In the examples above, the combustion of methane and gunpowder are exothermic, while the reduction of silver is endothermic. A reaction such as rusting will take place spontaneously in the presence of water even at relatively low temperatures, but many reactions, including exothermic ones like combustion, do not happen spontaneously. Instead they require an initial input of energy known as activation energy. Once this activation energy is supplied, an exothermic reaction like combustion of methane generates enough energy to keep itself going, and is said to be self-sustaining.

## Types of Reaction

There are many different types of reaction. The most straightforward is probably the combination reaction, which is where two or more reactants combine to produce a single product. A decomposition reaction is the opposite, where a single reactant breaks down to give two or more products.

A displacement reaction is where a more active element displaces a less active one from a compound. Metals in particular have a hierarchy of reactivity, with so-called alkali metals like sodium and magnesium at the top, followed by aluminum and zinc, with copper, silver, and gold at the bottom being the least reactive. Hence if you add zinc to a solution of silver salt dissolved in water, the zinc will displace the silver, which will precipitate out of the solution. But if you then add aluminum to the solution, the zinc will be displaced and precipitate out.

A combustion reaction is where a compound combines with oxygen—otherwise known as burning. Combustion is a particularly common example of a class of reactions known as "redox," short for reduction-oxidation. In a redox reaction electrons are swapped between reactants (see pp. 110–111); rusting is another example of a redox reaction.

## Balancing Out

An important principle of chemistry is that matter cannot be created or destroyed (except in nuclear reactions). This is known as the law of conservation of matter. In terms of reaction equations it means that there must be the same number of atoms on either side; new combinations are possible, but the overall output of atoms must equal the input. When writing an equation it is necessary to make sure it balances, and this is where scientific notation comes in so handy.

# CHEMISTRY IN THE MEDIEVAL ISLAMIC WORLD

The next stages in the evolution of chemistry took place in the Middle East, in the world of medieval Islam. During a golden age of intellectual activity a series of great names in alchemy formulated new principles, perfected techniques and processes, and preserved and built up a great store of knowledge that would profoundly influence the European Renaissance.

## The Silk Road and the House of Wisdom

Persia had long been a center of scholarship where influences and ideas from east and west, traveling the Silk Road alongside merchandise and people, met and merged. In late Antiquity the region received an influx of Classical knowledge from groups of Christians driven out of the Byzantine Empire by religious intolerance. Nestorian Christians, for instance, established the famous Persian medical school at Jundaishapur in around 500 CE.

The coming of Islam brought rapid and dramatic change to the region, with the Arab expansion of the 7th century bringing all of the Middle East and much of Central Asia, the Near East, and North Africa under Islamic rule. At first the Caliphate was hostile toward non-Islamic scholarship, but under the Abbasid Caliphate of the 8th–11th centuries the Islamic world saw a remarkable flowering of learning. The Translation Movement saw potentates actively hunting out the literature of ancient Greece so that it could be translated into Arabic, alongside important works from the eastern centers of ancient wisdom: India and China. Scholars flocked from across the

• *Mad dog biting a man* from a ca.1224 CE Abbasid Caliphate translation of a Greek *Materia Medica*. Islamic scholars preserved ancient learning and extended it with new ingredients, remedies, and techniques.

vast Islamic empire and beyond to the Abbasid capital at Baghdad, where institutions such as the famed Bayt al-Hikma (House of Wisdom) fostered the study of mathematics, astronomy, medicine, chemistry, zoology, geography, alchemy, and astrology. The rapid growth of scholarship was aided by high levels of literacy and the introduction of paper, a new technology brought from China along the Silk Road. A paper mill had been established in Baghdad by the 9th century CE.

## Greek Fire

While the collapse of the Roman Empire in Europe meant that the vast majority of ancient books were lost, the flame of Classical learning was kept alive and fanned to new heights in the Islamic world. Islamic alchemists drew on Classical sources such as Pythagoras, with his mystical mathematics, Aristotle and his elements, the physician Galen, whose physiological theories involving the four humors paralleled the Classical model of the elements, and the Neo-Platonists, whose mysticism and metaphysics provided the spiritual subtext for alchemy. Taking as well ideas from Chinese and Indian alchemy and science, a series of Islamic alchemists began systematically to explore and expand knowledge of substances and their reactions.

The first great name was Jabir ibn Hayyan (ca. 721–ca. 815), the "father of Islamic alchemy." He was succeeded by the Persian physician Al-Razi (ca. 865–ca. 925). Although closely affiliated with the work of Jabir, his writings are considered revolutionary because he began to approach a scientific philosophy. Moving away from the metaphysical baggage of his predecessor, Al-Razi was more willing to study substances in isolation, relying on his observations of what actually happened, rather than proceeding according to preconceived theories. Al-Razi's book *The Secret of Secrets* would become a bible for European alchemists; part of its significance was that it was set out almost like a laboratory manual, with a section describing all the exotic glassware that had been perfected by Islamic alchemists, and which would remain standard equipment in chemistry laboratories until the 19th century. In the last section of the book Al-Razi attempted to classify substances, arguably starting a process that would culminate in the periodic table. To the solid principles of sulfur and mercury adduced by Jabir (see pp. 44–45), he added a third, salt. This scheme would profoundly influence Paracelsus (see pp. 58–59).

After Al-Razi came Abu Ali ibn Sina (980–1037). Ibn Sina specialized in "iatro-alchemy"—alchemical medicine. He developed the Galenistic theory of the four humors, which mirrored the qualities or "natures" of Jabir and Aristotle before him. Ibn Sina was an important influence on the first wave of early Renaissance scholars in Europe, providing the authority that men like Roger Bacon and Albertus Magnus needed to pursue their own research into natural philosophy.

# Jabir ibn Hayyan

**Jabir ibn Hayyan was the first great figure in Islamic alchemy, although his achievements were not restricted to this field. His bold adaptation of Classical models of chemistry profoundly influenced later generations of alchemists; arguably of greater importance were the advances he made on the practical and experimental side, discovering new substances, mastering new techniques, and growing the body of chemical knowledge.**

## The Four Natures and the Two Metals

Jabir came to alchemy from the practical side, having practiced as an apothecary (the medieval equivalent of a dispensing chemist). It is interesting to note that apothecaries continued to drive the development of chemistry up to

the time of the French Enlightenment. He was no blinkered specialist, however; he took a holistic view of learning, seeing alchemy as just one aspect of natural philosophy.

Drawing inspiration from the legendary alchemical tome known as the *Emerald Tablet*, attributed to Hermes Trismegistus (see pp. 36–37), and from Aristotle's elemental theory, Jabir added new dimensions. He argued that Aristotle's elements were underlain by four corresponding "natures" (equivalent to qualities): hotness, coldness, dryness, and wetness. Pairs of these qualities created the four earthly elements (for example, hot + dry = fire).

He particularly focused on the nature of metals, which he believed were formed from two "metallic" elements that he identified with sulfur and mercury. The proportions of sulfur and mercury determined the nature of the metal; the perfect balance yielded gold. Like alchemists before and after him, Jabir was convinced he could find a way to transmute lead into gold—in his case, he believed that lead could be separated into its sulfur and mercury components, cleansed of impurities, and reconstituted in proportions that would produce gold. To achieve this it was necessary to use a substance that facilitated the transmutation but itself remained unchanged— what today is called a catalyst. In Hermes' *Hermetic Corpus* it was referred to as *xieron*, a dry or powdery substance; in Jabir's Arabic transliteration this became *al-iksir*, or "elixir."

• Jabir ibn Hayyan was a powerful force in both applied and theoretical chemistry. Here he is depicted with some of the specialist equipment he helped to develop.

## The Water of Kings

It was on the applied side that Jabir made the greatest contribution. He improved glass-making, the refining of metals, and fabrication of dyes and inks (for instance, developing an iron pyrite—fool's gold—ink as a cheap alternative to gold for use in illuminated manuscripts). He synthesized sal ammoniac (ammonium chloride), developed a technique for concentrating acetic acid, and invented a new acid, which came to be known as *aqua regia* ("royal water"). Now known to be a combination of hydrochloric and nitric acids, this new brew had the power to dissolve gold, which cannot be achieved with single acids alone.

He introduced organic substances to the alchemical mix by experimenting on plant extracts, although at the time philosophers saw no clear divide between the organic and inorganic worlds, viewing minerals, plants, and animals as part of a continuum of matter. In synthesizing new compounds, Jabir was seeking to discover and even create new species. Perhaps most valuably, Jabir recorded his experiments in a useful and systematic fashion, describing materials, equipment, techniques, and results. Not only did this mark a proto-scientific approach, it meant that his work could be a resource to later generations of alchemists.

• A sample of iron pyrite, the traditional name for iron (II) sulfide ($FeS_2$), popularly known as fool's gold because its brassy luster can be mistaken for the more valuable element.

## ARABIAN NIGHTS

Jabir led an eventful life against the backdrop of an eventful era, for he practiced during the reign of the caliph Harun al-Rashid, the legendary ruler of *Arabian Nights* fame. Born in Iran but of Arabian descent, he was immersed in the dangerous world of caliphate power-politics from the start. His father was executed for plotting to overthrow the Umayyad caliphate, and he himself was closely associated with Harun's vizier, Jafar, so that his fortunes rose and fell with his patron. When Jafar lost favor and was executed, Jabir was forced to flee the capital and retire to the country, where he occupied his last days penning his massive book *The Sum of Perfection*, one of hundreds of books attributed to him (although his authorship is often doubtful, as it was common practice to append a famous name to a work to make it more successful).

# CATALYSTS AND KINETICS

Jabir's work with catalysts opened an important new chapter in chemistry. To understand it we need to look first at kinetics—the study of reaction speeds. Kinetics examines the rate at which a reaction proceeds, and the factors that can affect this rate. Catalysts are one such factor. Temperature is another, affecting the speed of the reactive particles and therefore the rate at which they collide.

## Collision Theory

The simplest and most descriptive model of how reactions occur is known as collision theory. According to this model the atoms, and/or molecules, that constitute the reactants are like billiard balls, zooming around a table. For a reaction to occur they need to collide, and to do so with enough force to overcome the activation energy barrier (see diagram below). The moving particles have kinetic energy; if they have enough kinetic energy they can break chemical bonds and transfer their energy into new chemical bonds (which store it as chemical energy). As well as having sufficient energy, one reactive particle may also need to hit the other at just the right spot—each particle has a reactive site, and this is where the collision must occur.

## Turning Up the Heat

The temperature of a substance or mixture is a measure of the average kinetic energy of the particles making it up. Adding heat to a substance increases its temperature because the heat energy is converted into kinetic energy and the average kinetic energy of the particles goes up. In other words, a higher temperature means that particles have more energy, on average. Particles that have more energy are moving around more and are thus more likely to collide.

Another option for boosting reaction rates is to increase the concentration of the reactants. Essentially, having more reactive particles in a given volume of space increases the chances that they will collide, and the more collisions there are the more likely it is that a reaction will occur.

· If the collision between reactive particles has the required activation energy, they pass through a transition state in which they are at their highest energy state, before falling back down to the energy state of the products.

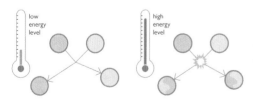

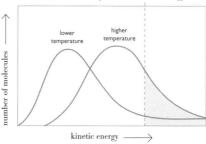

minimum kinetic energy to
provide activation energy

number of molecules →

lower
temperature

higher
temperature

kinetic energy ⟶

• Raising the temperature lifts more particles above the threshold minimum kinetic energy to provide the necessary activation energy, increasing the rate at which the reaction proceeds. In other words, raising the temperature of reactants is one way to boost reaction kinetics.

## Little Helpers

A catalyst is a substance that speeds up the rate of a chemical reaction but remains unaltered at the end of the reaction. Only a minuscule amount may be needed to produce a big effect. It is important to note that a catalyst does not increase the amount of end products, or alter the equilibrium of a reaction—to do so would be a violation of the laws of thermodynamics.

There are two types of catalyst—heterogeneous and homogeneous. A heterogeneous catalyst is one that is in a different phase from the reactants—typically it would be a solid (perhaps finely divided as a powder, or spread thinly over a wide surface area) while the reactants would be gases or liquids. Such a catalyst works by capturing one of the reactants and holding it in such a way as to present its reactive site, increasing the likelihood of another reactive particle colliding at the right spot and successfully reacting. This is how the platinum and palladium catalysts in your car exhaust's catalytic converter work.

A homogeneous catalyst is one that is in the same phase as the reactants. These catalysts often work by offering an alternative mechanism or reaction pathway for the reaction, one with a lower activation energy and faster kinetics. Typically the catalyst forms intermediate compounds at transition states in the new reaction pathway, before decoupling from the reactant and returning to its normal state.

For instance, if AB is the reactant, A and B are products, and C is the catalyst, the reaction pathway might look like this:

$$C + AB \longrightarrow CAB \longrightarrow CA + B \longrightarrow C + A + B$$

This is analogous to the much simpler equation: $C + AB \longrightarrow C + A + B$

• Addition of a catalyst changes the reaction pathway by forming intermediate products at lower energy levels, reducing the activation energy necessary to trigger the reaction.

ENERGY

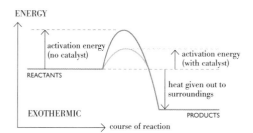

activation energy
(no catalyst)

activation energy
(with catalyst)

REACTANTS

heat given out to
surroundings

EXOTHERMIC

PRODUCTS

course of reaction

# **5** **Rate of a Chemical Reaction**

### THE PROBLEM:

Chemists define the speed, or rate, of a reaction as the amount of product made in a given time, or alternatively, the amount of starting material, reactant, used in that time. Understanding reaction rates is important across the chemical industry. For instance, agricultural fertilizers are made from ammonia ($NH_3$) gas, which is produced from nitrogen and hydrogen. Chemical engineer Amy has an order from a fertilizer manufacturer and needs to know how much ammonia she can supply. She is monitoring the rate at which hydrogen gas is being used up in her chemical plant—0.03 moles per liter per second—so at what rate is ammonia being produced?

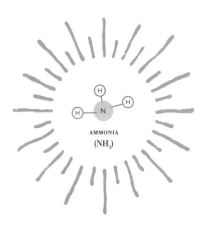

AMMONIA
($NH_3$)

### THE METHOD:

We can calculate the rate of ammonia gas produced by looking at how quickly the hydrogen gas is used. The mass of a mole (unit, mol) is the molecular mass of a substance in grams, and concentration is the number of moles per unit volume in units of mols per liter (mol $L^{-1}$). The rate is the change in concentration over time expressed in units of moles per liter per second (mol $L^{-1}$ $s^{-1}$).

## THE SOLUTION:

We know that the rate of loss (or use) of hydrogen gas is 0.03 mol $L^{-1}$ $s^{-1}$ for the following reaction:

$$N_2(g) + 3 H_2(g) \longrightarrow 2 NH_3(g)$$
$$1 \qquad 3 \qquad \qquad 2 \text{ moles}$$

Looking at the equation, we see that for every 3 moles hydrogen ($H_2$) used there are 2 moles ammonia ($NH_3$) produced. To calculate the rate of ammonia production, we need to work out how much ammonia is produced per mole of hydrogen, and then multiply this by the rate at which hydrogen is being used in Amy's laboratory:

$$(2 / 3) \text{ x } 0.03 = 0.02 \text{ mol } L^{-1} s^{-1}.$$

Chemical reactions proceed at different rates and the rate of any particular reaction depends on several factors; including the physical state, concentration and pressure (if gases) of the reactants, the temperature and whether a catalyst is used. A catalyst is a substance that increases the rate of a reaction but remains chemically unchanged at the end of the reaction. This reaction, known as the Haber-Bosch process, works best with an iron catalyst at a pressure of 150–250 atmospheres (atm) and at a temperature of 550–1,000°F (300–550°C). The experimental study of reaction rates is termed chemical kinetics. For gas reactions, partial pressures are more often used than using terms of concentration.

• Between 1894 and 1911 the German chemist Fritz Haber successfully developed a method of forming ammonia from naturally occurring nitrogen products such as sodium nitrate. The discovery of this process, known as the Haber-Bosch, led to Haber being awarded the 1918 Nobel Prize in Chemistry.

# POISONS AND POISONERS

One of Jabir's many works was the *Kitab al-sumum* (*Book of Poisons*), a seminal work in toxicology (the study of poisons). Jabir and other Islamic proto-chemists were instrumental in passing on to medieval Europe the vast body of lore and practice that had built up around the use and treatment of poisons, one of the darker and less savory aspects of chemistry.

## From Otzi to the Borgias

Mankind's relationship with chemicals has always been a two-way street—not only have humans long made use of chemicals, they have also been affected by them. When a chemical is injurious to the health of a biological system even when present in tiny quantities, it is said to be a toxin (a word that itself derives from the Greek *toxicon*, which described the poison used on arrow tips). Chemicals from plant, animal, and mineral sources have been known and used as poisons for millennia. The earliest known case of poisoning is probably that of Otzi the Iceman, whose frozen body was found in an Italian glacier 5,000 years after his death. Analysis of his hair showed that he had been suffering from chronic arsenic poisoning, probably as a result of smelting copper ores contaminated with the toxic element.

• The alchemical symbol for red arsenic (realgar), an arsenic sulfide ($As_4S_4$).

Arsenic was not isolated as an element until 1250, when the alchemist Albertus Magnus (ca. 1200–1280) heated white arsenic with soap, but it had been known through its ores, realgar and orpiment, for thousands of years before this. The potency of arsenical compounds as poisons and medicines was known to the ancient Romans, Indians, and Chinese. The Romans knew how to make the highly soluble and thus highly toxic salt sodium arsenite, by heating white arsenic with natron salt. Indeed there was an epidemic of arsenic murders in ancient Rome that prompted Ovid to lament that, "The husband longed for the death of his wife, she of her husband; murderous stepmothers brewed deadly poisons, and sons inquired into their fathers' years before the time."

But the golden age of poisoning, sometimes known as the Age of Arsenic, was Renaissance Italy. This was a historical moment when Classical learning was being rediscovered and extended, and the alchemist's expertise with arsenical compounds, particularly white arsenic or arsenic trioxide, a flavorless and odorless white powder that could be added to food or drink, was extensively misappropriated. The most notorious poisoners of the late 15th century were the Borgias, infamous for inviting enemies to dinner and feeding them arsenic-laced dishes. Supposedly, Pope Alexander and his son Cesare Borgia were victims of their own game, however, accidentally drinking their own poisoned wine.

• Cesare Borgia, who reputedly perished after mistakenly drinking his own poisoned wine.

## Murder for Sale

Over the next two centuries poisoning became wildly fashionable. In 17th-century Rome a group of young wives were supplied with a "clear, tasteless, and limpid" arsenic-based brew known as *Aquetta di Perugia*, which they obtained from Hieronyma Spara, described by Victorian historian Charles Mackay as a "hag, reputed witch, and fortune-teller." Spara was eventually tortured and executed, but soon after, another mistress of poisons, Tophania of Naples, started selling *Aqua Toffana*, of which just four drops were needed to dispose of inconvenient relatives. Such poisons were known in France as *poudres de succession*, because they cleared the way for poisoners to come into their inheritances early.

While the concoction and supply of toxins gave early chemists a bad name, the study of poisons also stimulated some important advances in the field. Physicians and alchemists such as Paracelsus (see pp. 58–59), Ellenborg, and Agricola linked specific chemicals to specific effects on the body, and elucidated important principles of toxicology such as the relationship between the amount of a chemical and the body's response to it. Through manufacturing, isolating, and analyzing poisons, Renaissance proto-scientists helped lay the foundations for analytical chemistry, the branch of chemical science concerned with detecting and identifying chemicals.

### VINEGAR CURE

Ancient Sumerian and Akkadian texts recorded knowledge of poisons and emphasized the use of vinegar to detoxify poisons, a practice that continued for thousands of years. We now know that acetic acid in vinegar can effectively attack the chemical bonds that hold poison molecules together, breaking them up into less toxic components.

# RENAISSANCE ALCHEMY

As Classical learning filtered into Europe via the Islamic scholars, alchemy became the essential pursuit for men who wanted to understand the hidden workings of the universe. European scholars of the late medieval and early modern period were driven by the quest to read what they called the Book of Nature; alchemy appeared to them to be the key to deciphering this great wisdom.

## The Quest for the Philosopher's Stone

Alchemy can be understood on many levels. The most obvious of its goals was the transmutation of base metals into gold, which alchemists hoped to achieve using a mythical substance known as the Philosopher's Stone. Essentially a sort of magical catalyst, the Philosopher's Stone had many powers attributed to it. According to Arnold of Villanova (ca. 1238–ca. 1310): "There exists in Nature a certain pure substance, which when discovered and brought by Art to its perfect state, will convert to perfection all imperfect bodies that it touches."

Alchemical tomes such as the *Emerald Tablet* described in dense allegory and arcane symbolism procedures for making the Philosopher's Stone, and modern science generally regards the quest for this magical substance as a deluded fantasy pursued by mountebanks and charlatans. It is certainly true that many of those who practiced alchemy were fools motivated by greed, or con artists trying to cheat money from unwise patrons; known as "puffers" or "mere chymists," these men gave alchemy such a bad name that it was periodically prohibited by kings and popes.

Yet the greatest minds of the era also pursued the alchemist's art. As historian of science Paul Strathern points out, it was at the time "the only real science of matter." Alchemy seemed to offer a rational system for discovering the secrets of nature, and what seems magical to us was in fact considered a form of technology—the application of knowledge to allow control of nature.

• Joseph Wright of Derby's famous 1771 painting, *The Alchymist, In Search of the Philosopher's Stone*, depicts a fanciful, medievalized version of the discovery of phosphorus by 17th-century alchemist Hennig Brand.

The philosophy that gave impetus to the pursuit of alchemy was essentially the same one that created science; the belief that the systematic interrogation of nature through experiment will allow man to discover her secrets and become her master. As the 19th-century chemist Justus von Liebig observed, "Without the Philosopher's Stone, chemistry would not be what it is today. In order to discover that no such thing existed, it was necessary to ransack and analyze every substance known on Earth."

## The Elixir of Life

The pursuit of this elusive quarry led Renaissance alchemists to make many important discoveries, expanding the "toolbox" available to later natural philosophers through adding to the list of known chemicals. In 1250, for instance, the clerical scholar Albertus Magnus isolated elemental arsenic, while his student Roger Bacon (ca. 1214–1292) may have invented black powder (a form of gunpowder) independently of Chinese influence. Bacon also championed the cause of experimentation as the best route for uncovering the truth about nature, inspiring later scholars such as Robert Boyle (see pp. 74–75).

A pseudonymous 14th-century author who styled himself after Geber (as Jabir was known in Europe), and hence is known as the False Geber, made important discoveries such as vitriol and *aqua fortis*. Vitriol is sulfuric acid, a vital tool in chemical analysis,

(see pp. 58–59)

described as the most significant chemical advance since the discovery of iron smelting, while aqua fortis is strong nitric acid. These tools would later allow the isolation of individual elements from compounds.

Meanwhile the search for the Elixir of Life, another mythical substance supposed to cure all ills and confer immortality, led to important breakthroughs in pharmacological alchemy, such as the discoveries of Paracelsus (see pp. 58–59). Arnold of Villanova, believing that grapes absorbed the essence of the Sun and therefore of gold, distilled wine to produce *aqua vitae*— practically pure alcohol, another important tool for later chemists. Alcohol, like the strong acids, can act as a solvent for substances that do not dissolve in water.

---

**ALCHEMY AND THE DRINKS CABINET**

Many traditional tipples, from whisky to obscure liquors produced by monasteries, owe their origins to the alchemist's search for the Elixir of Life, and specifically to Arnold of Villanova's distillation of *aqua vitae*. A classic example is Chartreuse, made by Carthusian monks, supposedly from a recipe in an alchemical manuscript titled *An Elixir of Long Life*, presented to them in 1605.

# SOLVENTS AND SOLUTIONS

For the alchemists the process of dissolving a substance seemed almost magical; by using the right liquid, almost any substance could be made to disappear—apparently—and then through evaporation, condensation, or precipitation the original substance or a new one could be recovered. The core process is now better understood, constituting one of the basic concepts of chemistry.

## What Is a Solution?

A solution is defined as a homogeneous mixture; in other words, a mixture that is the same throughout, so that if you took a sample from the top and another from the bottom they would have the exact same makeup. This is different from a suspension, in which particles of one substance float around in another and can be filtered out. A solution is made up of a solvent and one or more solutes. Generally the substance that is in the majority is said to be the solvent.

Generally solvents are liquids while solutes can be in any phase, but there can also be gaseous and solid solutions. Gases can mix to become homogeneous; the air is a good example—sample it anywhere at sea level and it will have the same gases in the same proportions. The majority of the atmosphere is made up of nitrogen, so this is considered the solvent, while oxygen, carbon dioxide, and so on are solutes. Solid solutions include metal alloys—bronze, for instance, is a solid solution of tin dissolved in copper solvent.

## Like Dissolves Like

Water is the best known and most common solvent, but not everything will dissolve in water. The principle that determines solubility is "like dissolves like," where "like" is a reference to polarity. Polarity is an electrical property present in some molecules, caused by the type of bonds between the atoms. In a molecule like water, where two hydrogen atoms are bonded to an oxygen atom, the electrons are unevenly distributed so that the oxygen atom has a partial negative charge and the hydrogen atoms have a partial positive charge. The molecule itself thus ends up with a negative pole and a positive pole, like a tiny magnet, and is known as a dipole. Because of its polar nature, water can dissolve other polar solutes, such as salts, sugars, and alcohols, but nonpolar solutes, such as oils, will not dissolve in water. Instead they will dissolve in non-polar solvents; for instance, olive oil will dissolve in petroleum.

## THE TYNDALL EFFECT

Colloids are mixtures where the solute particle size is between that of a solution and a suspension (generally around 1–1000 nanometers). At this scale the particles are not a true solution, but are small enough not to settle as they would in a suspension, and may be invisible. One way to determine if a mixture is a colloid or a solution is the Tyndall effect, in which a beam of light passed through a colloid is visible, but invisible when passed through a solution.

• John Tyndall (1820–1893) was a prolific author across a wide range of scientific subjects, from glaciology to experimental physics.

## Solubility and Saturation

The maximum amount of solute that will dissolve in a solvent is described as its solubility. Solubility is usually measured in grams of solute per 100 ml of solvent (g/100ml). For solids, solubility increases with temperature, so that, for instance, you can dissolve more sugar in a cup of hot tea than a glass of iced tea. For gases dissolving in liquids this relationship is reversed—as temperature rises, less gas will dissolve. When the maximum amount of solute is dissolved the solution is said to be saturated. Sometimes it is possible to dissolve more than the theoretical maximum, in which case the solution is said to be supersaturated. Something as simple as shaking a supersaturated solution is often enough to cause the excess solute to precipitate out of solution.

The concentration of a solution is a measure of how much solute is dissolved in it; it can be expressed in a number of different ways. Units of concentration include molarity (number of moles of solute per liter of solution—see pp. 48–49 for an explanation of moles); parts-per-million (or billion)—commonly used for gaseous solutions; and percentage. This last measure can itself be defined by weight, volume, or a combination of both. For instance, in 100 grams of a salt solution that has a concentration of 10% by weight, there will be 10 grams of salt. Concentrations of alcoholic drinks are given in percentage by volume, so that in 1 liter of wine that has an alcohol concentration of 12% by volume there will be 120 ml of alcohol.

# Water Solubility

## THE PROBLEM:

Barium sulfate is used in medicine for radio-imaging of the gastrointestinal tract. While all water-soluble barium salts are highly toxic, ingestion of the insoluble barium sulfate (as a "barium meal") for examination is safe. It is also used in industry as a white pigment in paint. Stacey works at a chemical supplies company offering barium sulfate ($BaSO_4$) to hospitals. Using a solution containing 50 grams of sodium sulfate, how does Stacey work out how much barium sulfate she can make?

## THE METHOD:

Barium sulfate can be produced from the reaction of barium chloride solution and a solution of a sulfate (here, sodium sulfate, $Na_2SO_4$). It precipitates out of solution as crystals of an insoluble white solid. Using the atomic weights of all the elements involved, Stacey can calculate the molecular masses of the starting materials and the products for her reaction. The equation for the reaction of barium chloride with sodium sulfate is:

$$BaCl_2(aq) + Na_2SO_4(aq) \longrightarrow BaSO_4(s) + 2\,NaCl(aq)$$

## THE SOLUTION:

The atomic weights for the elements in the equation are: barium (Ba = 137); chlorine (Cl = 35.5); sodium (Na = 23); sulfur (S = 32); and oxygen (O = 16). Using these, we can calculate the relative molecular masses of the compounds:

Barium chloride ($BaCl_2$):
$$37 + (2 \times 35.5) = 208$$

Sodium sulfate ($Na_2SO_4$):
$$(23 \times 2) + 32 + (16 \times 4) = 142$$

Barium sulfate ($BaSO_4$):
$$137 + 32 + (4 \times 16) = 233$$

Sodium chloride (NaCl)
$$23 + 35.5 = 58.5$$

Now that we have the molecular masses for all four compounds, we can plug these into the equation given on the previous page:

$$BaCl_2(aq) + Na_2SO_4(aq) \longrightarrow BaSO_4(s) + 2\,NaCl(aq)$$
$$\quad 208 \qquad\quad 142 \qquad\qquad 233 \qquad\quad 117$$

As the equation shows, 208 grams of $BaCl_2$ reacts with 142 grams of $Na_2SO_4$ to give 233 grams of $BaSO_4$. Given that Stacey is using a solution that contains 50 grams, rather than 142 grams, of sodium sulfate, the corresponding value for $BaCl_2$ must be calculated:

$$(208 / 142) \times 50 = 73.23 \text{ grams of } BaCl_2$$

In doing the same calculation, but this time with $BaSO_4$, we can find the amount of barium sulfate produced:

$$(233 / 142) \times 50 = 82 \text{ grams of } BaSO_4$$

The amount of barium sulfate that Stacey can produce is therefore 82 grams.

• An X-ray of the bowels of a patient after ingestion of barium sulfate. Barium is radio-opaque and is often used alongside radiolucent compounds such as carbon dioxide to highlight abnormalities in the gut.

# Paracelsus

**One of the more colorful and controversial characters of his era, the 15th-century physician and alchemist known as Paracelsus made significant advances in chemistry and especially in the chemistry of medicine. As importantly, his bombastic manner helped free natural philosophers from the chains of tradition, opening doors to new methods of enquiry and scientific exploration.**

## The Wandering Scholar

Philippus Aureolus Theophrastus Bombastus von Hohenheim (1493–1541) was born in Switzerland, the son of a doctor. Educated at a school that specialized in training engineers for the nearby silver mines, he gained a firm grounding in mineralogy and metallurgy, and also in the practical approach found in industry (in contrast to the purely scholastic approach current in the universities). After this he became an itinerant scholar. Sources differ on the extent of his wanderings, but after studying (and possibly gaining degrees in) medicine at the Universities of Vienna, Basel, and Ferrara among others, he became a military surgeon and traveled at least as far as Constantinople. According to legend he trained with mystics, alchemists, and physicians in Egypt, Arabia, and the Holy Land, and may have supported himself through astrology and other occult pursuits. Among the many accusations later leveled at him was the charge of necromancy.

In 1526 he returned to Basel and promptly made a name for himself by curing the famous printer Froben of a leg infection, where other physicians had advised amputation. On the basis of this he was appointed city physician, although he did not keep the post long. Repeatedly throughout his career he managed to antagonize the local authorities and make himself unwelcome. He made a bad start by publicly burning the works of Galen and Avicenna and adopting the name Paracelsus, thus declaring that he had moved "beyond Celsus" (an important Roman medical encyclopedist of the 1st century CE). After many years traveling in Europe and moving between positions he returned in 1541 to his home town of Villach and was appointed physician to the Duke of Bavaria, dying that same year.

## A New Spirit of Investigation

Paracelsus' reputation rests chiefly on his bold new approach to medicine, which prefigured many aspects of modern doctoring, most notably what historian Hugh Trevor-Roper calls his "insertion of chemistry into medicine." Paracelsus understood that specific chemicals have specific effects, and just as importantly, that these effects depend on dose.

"Alchemy is the art that separates what is useful from what is not by transforming it into its ultimate matter and essence . . . [it] is an explanation of the properties of all the four elements, that is to say, of the whole cosmos."—*Paracelsus*

## THE HOMUNCULUS

One of Paracelsus' more unusual claims was that he had successfully created a *homunculus*, or "little man" (an artificial being, precursor of the golem and Frankenstein's monster). According to the recipe he gave, if sperm, "enclosed in a hermetically sealed glass, is buried in horse manure for forty days, and properly magnetized, it begins to live and move. After such a time it bears the form and resemblance of a human being, but it will be transparent and without a body." Feeding an extract of blood he called *Arcanum sanguinis hominis* would result in a tiny human child that could be raised as normal.

• An illustration from Goethe's *Faust*, showing Wagner, former student of the eponymous hero, in his alchemical laboratory, laboring to create a homunculus.

He originated mercury treatments for syphilis, described the anesthetic ether and concocted laudanum (a tincture of opium in alcohol) as a general pain-reliever and cure-all.

More generally he is credited with helping to set alchemy on the road to becoming chemistry. He formulated a new theory of matter, arguing that the basic principles from which all substances were composed were the *tria prima* of sulfur, mercury, and salt, representing the qualities of combustibility, liquidity, and solidity. He discovered new compounds from his study of the chemistry of mercury, zinc, cobalt, bismuth, potassium, antimony, and other metals, and was the first to concentrate alcohol by freezing it. Paracelsus even devised proto-scientific nomenclature for chemicals, and was one of the first to attempt a systematic classification of substances according to their chemistry.

But perhaps his true legacy to later philosophers was his willingness to challenge received authority. The dead hand of scholasticism (the system of enquiry based on Aristotle, deductive logic and the primacy of authority, particularly that of the Church) retarded the development of natural philosophy; with his brash assault on its sacred cows, Paracelsus opened the way for the next generation. Historian of science Charles Webster argues that Paracelsus' approach was integral to "a new spirit of investigation," which inspired such figures as Francis Bacon, Thomas Boyle, and the founders of the Royal Society.

# WEIGHTS AND MEASURES

Paracelsus famously said, "All substances are poisons, there is none that is not a poison. Only the dose determines that a thing is not a poison." In identifying what modern toxicologists call the dose-response relationship, he was among the first to recognize what would become one of the most important, if least glamorous, of the elements of the new chemistry that was about to dawn: the vital need to measure things properly.

## The Principle Clue

Aristotelian natural philosophy was largely qualitative in nature—that is, it discussed categories, classifications, qualities, and essences. Although alchemy introduced some elements of quantitative thinking, such as occasional directions on quantities in the recipes given in alchemical manuscripts, it too remained largely qualitative. The precise amounts of substances used by alchemists were less important than their nature. But chemistry is a profoundly quantitative science, particularly in regards to the search for new compounds and elements. Without precise measurements of reactants, there was no way for the coming chemists to

---

• **BLESSED ARE THE INSTRUMENT MAKERS**

*According to one school of thought the crucible of modern science was Louvain, in modern-day Belgium. It was here that the mathematician and astronomer Gemma Frisius (1508–1555) and his pupil Gerard Mercator (1512–1594) started making scientific, surveying, and mapmaking instruments. While the academic establishment still sneered at vulgar "artificers and mechanicks," the instruments of Louvain allowed natural philosophers to measure the world as it actually was, rather than as the ancient texts claimed. Forging links with booksellers and merchants, the Louvain workshops started an international trade that would bear fruit in telescopes and microscopes, balances, and scales—literally the instruments of the Scientific Revolution.*

• Gemma Frisius, a mathematician who app[lied] his expertise to cartography and became [one] of the leading mapmakers of the age.

understand properly the nature of the products. The need for a new approach based on measurement can be traced back to a few important figures in the history of science.

Nicholas of Cusa (1401–1464) was a theologian and natural philosopher who followed Plato in considering the material world to be a shadow of the ideal world. Accordingly he argued that it was only through the language of mathematics that the true nature of things could be perceived: "Number is the principle clue which leads to wisdom." Nicholas applied this principle to experiments. He weighed a ball of wool at different times and noted that it changed according to how much water was absorbed from the atmosphere, so that it became an instrument for measuring atmospheric humidity. He used weighing to measure the volume of water in circular and square containers, enabling him to estimate pi to a high degree of accuracy. Most famously, he weighed a plant growing in a pot, with a degree of accuracy not attempted previously, allowing him to show that it was gaining weight, if only by minuscule amounts. Here was the first inkling that plants could be taking something from the air, and that the air itself had weight.

"Man, being the servant and interpreter of Nature, can do and understand so much only as he has observed."—*Francis Bacon*

## A New Pattern of the World

The theme of measurement was also stressed by Galileo (1564–1642), who distinguished between primary qualities that could be measured objectively and were thus accessible to experimentation, as opposed to secondary qualities that were perceived subjectively. This distinction, between what could only be subjectively known and what could be shown to be objectively true by measurement and experiment, was essential to the emerging scientific philosophy (see pp. 82–83 on the scientific method). The Elizabethan statesman and philosopher Sir Francis Bacon (1561–1626) warned, "God forbid we should give out a dream of our own imagination for a pattern of the world."

# 7 Atomic Weights

## THE PROBLEM:

Chuck is examining rock salt (sodium chloride, NaCl) in his science lab and wants to know its molecular weight ($M_r$). Looking at a periodic table on the wall, he wants to know why the element chlorine has an atomic weight of 35.453. He asks his science teacher who explains that the element exists in nature in the form of two isotopes, chlorine-35 and chlorine-37, which are abundant in a 3:1 ratio. The teacher decides, as an assignment, to ask the whole class how the atomic weight of chlorine (35.453) is obtained. How do they carry this out?

## THE METHOD:

In order to answer this question, it's worth going over some of the key terms:

The *atomic weight* ($A_r$) of an element is an average of the masses of its isotopes, weighted to take into account each isotope's natural abundance. An element's *isotopes* are atoms of that element that contain a different numbers of neutrons (see Exercise #16, pp. 124–125). For example, chlorine has two isotopes:

chlorine-35, which contains 18 neutrons; and chlorine-37, which contains 20 neutrons. The atomic weight of chlorine is therefore an average of these two.

The *natural abundance* (NA) of an isotope is a figure, expressed as a percentage, that shows how common it is in relation to the other isotopes of the same element.

Knowing that the atomic weight of an element depends on the mass and natural abundance of its isotopes, the students in the class calculate the chlorine's atomic weight by adding the mass of each isotope multiplied by its natural abundance. For Cl, which has isotopes $^{35}$Cl and $^{37}$Cl, the equation the class must solve to establish the correct atomic weight is:

$$\text{(Mass of } ^{35}\text{Cl x Abundance of } ^{35}\text{Cl)}$$
$$+$$
$$\text{(Mass of } ^{37}\text{Cl x Abundance of } ^{37}\text{Cl)}$$

## THE SOLUTION:

From official tables, they discover that:

$^{35}$Cl isotope has a mass of 34.968853 and a natural abundance of 24.24% (0.2424)

$^{37}$Cl has a mass of 36.965903 and a natural abundance of 75.76% (0.7576)

Plugging these figures into the equation above, we can work out the atomic weight of chlorine:

$$(36.965903 \times 0.2424) +$$
$$(34.968853 \times 0.7576) = 35.453.$$

Here we have, then, the atomic weight of chlorine as it is displayed on the periodic table. Try using this formula to calculate the atomic weight of another element.

• Sodium chloride forms a giant (indefinitely repeating) lattice typical of ionic solids. Each atom has six neighbors, so the crystal is described as 6:6 coordinated.

○ Na

○ Cl

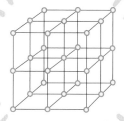

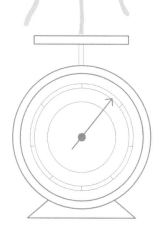

# PNEUMATIC CHEMISTRY

With his experimental look at air, Nicholas of Cusa had opened a whole new chapter in chemistry—the study of pneumatics (from the Greek *pneuma*, "breath"). Although alchemists had been aware of vapors and airs arising from their crucibles and flasks, little attention had been paid to the world of gases. This was to change, with major consequences for the new science of chemistry.

## Entitled to the Denomination

As late as 1727, the English clergyman and scientist Stephen Hales (1677–1761), in his *Vegetable Staticks* of 1727, a pioneering work on pneumatics, was moved to ask: "May we not with good reason adopt this now fixt, now volatile Proteus [the gas phase of matter] among the Chymical principles . . . notwith-standing it has hitherto been overlooked and rejected by Chymists, as in no way entitled to the Denomination?" Gases were the poor relations of the chemical world, ignored and misidentified by alchemists and natural philosophers, who, if they thought about it at all, generally assumed that all "airs" were one and the same—ethereal, insubstan-tial, and unknowable.

But in the 17th century this began to change, first with the work of Jan Baptista van Helmont (see pp. 66–67), and then with a series of dramatic experiments proving the existence of the vacuum. The first of these was the barometric experiment of Evangelista Torricelli (1608–1647), performed in 1644. Torricelli filled a glass tube with mercury. Then he put his finger over the open end, turned it upside down and lowered the mouth into a basin of mercury. When he took his finger away the column of mercury inside the tube descended part of the way and then stopped. The space at the top of the tube was now empty, Torricelli insisted—it was a true vacuum, contrary to the claims of the ancients, who had insisted that Nature abhorred a vacuum. What was more, he pointed out, the column of mercury was being held in place by the pressure of the air on the basin of mercury; in other words, air had weight.

## Ideal Gas Laws

What these experiments showed was that air had mass while a vacuum was true empty space. Natural philosophers like Boyle (see pp. 74–75) saw this as clear evidence for the atomic theory of matter, with substances made up of tiny particles separated by void. Eventually such thinking would give rise to a model of how gases behave, known as the kinetic molecular theory of gases, which describes the properties of an ideal gas.

These include:

- Gases are composed of minute particles. It makes no difference whether these are atoms or molecules, they behave the same. The particles are considered to be so small in relation to the distances between them that they effectively take up no volume, which means they can be compressed (unlike a liquid or solid).

- The gas particles move about randomly, in straight lines, until they collide with the walls of their container. These collisions constitute the pressure exerted by a gas. This constant, random, uniform motion allows gases to mix uniformly.
- Gas particles have negligible forces of attraction or repulsion between them, so they can be considered as completely independent, like minute ball bearings careering around.
- The average kinetic energy of the gas particles determines the temperature of the gas.

## • THE MAGDEBURG SPHERE

*An even more dramatic exposition was to follow. In 1650 German military engineer and mayor of Magdeburg, Otto von Guericke (1602–1686), modified a water pump to create the first air pump, which he used to pump the air out of containers to create vacuums, proving that inside a vacuum sound would not travel, wood would not burn, and animals could not breathe. In 1654 he performed his most famous demonstration in front of the emperor Ferdinand III. Two giant copper hemispheres were fitted together and the air pumped out of them. Although no screws or ties held them*

*together, the hemispheres could not be pulled apart by 16 horses; when Guericke released a valve to re-admit the air, the hemispheres fell apart by themselves.*

• Engraving of the Magdeburg sphere demonstration, showing the two teams of horses failing to separate the copper hemispheres stuck together with a vacuum.

# Jan Baptista van Helmont

**The father of pneumatic chemistry was a reclusive Flemish nobleman with a mystical bent and arcane beliefs, yet who is credited with the first controlled experiment in biochemistry and who anticipated important developments and laws in chemistry. His groundbreaking experiments led him to conclusions both millennia out of date and centuries ahead of his time.**

## Reclusive Researcher

Jan Baptista van Helmont (1579–1644) was a Flemish physician and alchemist from a noble family. After studying at Louvain University and traveling around Europe he retired to his country estate to pursue mystical and scientific researches. Despite being very religious he ran into trouble with the Catholic Church over his involvement in a controversy over the powder of sympathy—the belief that a special ointment could treat wounds through application to the blade that had caused them (see box). For insisting this was a purely natural phenomenon with nothing magical about it, Van Helmont was apprehended by the authorities and put under house arrest. Not until after his death was his son able to publish his collected writings as the 1648 volume, *Ortus Medicinae* (*Origin of Medicine*).

## The Womb of the Waters

Van Helmont's most famous research was an improved version of Nicholas of Cusa's 15th-century experiment, which he described concisely in his posthumous book: "For I took an Earthen Vessel, in which I put 200 pounds of Earth that had been dried in a Furnace … and I implanted therein the … Stem of a Willow Tree, weighing five pounds and about three ounces …" He covered the vessel with a tin shield and fed the plant only distilled water, and after five years he dug up the small tree, found that it weighed 169 pounds, and once again dried and weighed the earth in the pot, "and there were found the same 200 pounds, wanting about two ounces. Therefore 164 pounds of Wood, Barks and Roots arose out of water only."

At the time the mechanism of photosynthesis was obviously unknown, so Van Helmont's natural conclusion was that the tree had transformed water into wood, bark, and leaves. His alchemical research had already shown that "substantial bodies" could be reduced to water by dissolving them in acid, all of which he saw as proof of his belief that water was the chief constituent of matter: "The whole rank of Minerals, do find their Seeds in the Matrix or Womb of the Waters …" (just as Thales had originally claimed 2,000 years earlier).

• Van Helmont refused to accept his initial degree from Louvain University, claiming he had learned nothing.

## THE POWDER OF SYMPATHY

One of the magical principles of alchemy is the power of "sympathy": the belief that objects or substances, once associated, retain influence over one another. This principle informed a peculiar belief of Paracelsus' later taken up by Van Helmont, that a wound could be treated by applying a special ointment (later known as the "powder of sympathy") to the blade that had inflicted it. The ingredients included moss from the skull of a man who had died a violent death, boar and bear fat from animals killed while mating, burnt worms, dried boar's brain, red sandalwood, and powdered mummy. According to Sir Francis Bacon, "it may be applied to the Weapon, though the Party Hurt be at a great Distance [and what's more] it seemeth the Imagination of the Party, to be Cured, is not needful! to Concurre; For it may be done without the knowledge of the Party Wounded ..."

## Spirits of the Air

Although he had overlooked the role of carbon dioxide in plant growth, he nonetheless became the first man to postulate its existence, after another groundbreaking experiment involving careful weighing. He set fire to 62 pounds (28 kilograms) of charcoal and found that only 1 pound (50 grams) of ash was left. Van Helmont believed that matter could not be destroyed, only changed in form (anticipating by over a century the law of conservation of mass), having previously demonstrated that metal dissolved in acid can be recovered without loss of weight. Accordingly he surmised that the other 61 pounds of matter had escaped as a form of vapor or airy spirit, which he named after the Greek word for chaos: "I call this Spirit, unknown hitherto, by the new name of Gas, which can neither be constrained by Vessels, nor reduced into a visible body."

The gas from burning charcoal he specifically named spiritus sylvester ("spirit of the wood"), also identifying from other experiments in combustion a second form of spiritus sylvester, gas carbonum and gas pingue, the four of which we now call carbon dioxide, carbon monoxide, nitrous oxide, and methane.

# The Air That We Breathe

## THE PROBLEM:

Carbon dioxide is essential to life on Earth; plants absorb it from the air and use the Sun's energy to make sugars during photosynthesis, releasing oxygen as a waste product. Unfortunately, it is also a "greenhouse gas" that is released by burning fossil fuels, and rising levels of it in the atmosphere are responsible for global climate change. Maxine is worried about her next road trip and would like to figure out how much $CO_2$ she will add to the atmosphere when her car uses 50 liters of gasoline.

## THE METHOD:

Earth's atmosphere comprises four main gases: nitrogen (78.084%), oxygen (20.948%), argon (0.934%), and carbon dioxide (0.0314%), with trace amounts of other gases and 1–5% water vapor also present. Gasoline contains a mixture of hydrocarbons, mainly isooctane and heptane, hexane, and pentanes, which are made of hydrogen and carbon, hence the name hydrocarbon.

For simplicity, we can assume gasoline is mostly isooctane ($C_8H_{18}$). The simplified chemical reaction taking place in the engine of Maxine's car is the burning of isooctane in air (oxygen):

$$2\ C_8H_{18} + 25\ O_2 \longrightarrow 16\ CO_2 + 18\ H_2O$$

Although this equation uses two molecules of isooctane ($C_8H_{18}$), for the sake of simplicity we will refer to one molecule from now on. This means that eight molecules of carbon dioxide will be produced, rather than the 16 shown in the equation.

## THE SOLUTION:

For each isooctane molecule burned, eight molecules of $CO_2$ are produced. Using $A_r$ for carbon ($C = 12$), hydrogen ($H = 1$) and oxygen ($O = 16$), we can calculate 1 mole of isooctane ($C_8H_{18}$) weighs 114 grams and 1 mole of $CO_2$ gas weighs 44 grams.

Now that we have the weight in grams of 1 mole of isooctane and carbon dioxide, the next step is to work out how many grams of isooctane, and thus of carbon dioxide, there are in 50 liters of gasoline. Isooctane has a density of 0.6919 grams per milliliter, so 1 liter of gasoline will contain 691.9 grams of isooctane. By converting this into moles, we can determine the amount of carbon dioxide contained within this liter of fuel:

$691.9 / 114 = 6.07$ moles of isooctane

Referring back to the equation on the previous page, we can calculate the amount of $CO_2$ this contains:

6.07 x 8 (molecules) x 44 (weight in grams) = 2137 grams, or 2.14 kilograms of $CO_2$.

Therefore, every 50 liters of gasoline used will produce 107 kilograms (2.14 x 50) of $CO_2$.

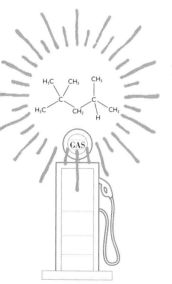

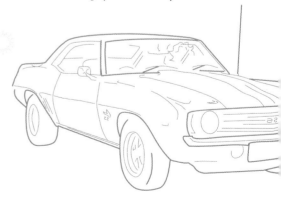

• Isooctane, the main hydrocarbon in gasoline, has eight carbon atoms. With this many carbon atoms there are a large number of potential arrangements, or isomers, each of which has a slightly different chemistry.

# ACIDS AND BASES

One of the oldest distinctions in chemistry was between acids and
alkalis: the former with sharp, sour tastes and the ability to "dissolve"
earths and metals; the latter with a bitter flavor and often a soapy feel.
A third class of substance was known as salts, but by the 17th century it
was becoming clear that salts could be the product of "antagonism"
between acids and alkalis, and substances that reacted with acids to
form salts (including metals) were called "bases."

## The Fabric of Chemistry

Acids and their "antagonists" the bases
were to play a vital role in the emer-
gence of chemistry as a science, so
much so that the early 19th-century
philosopher and historian of science
William Whewell (the man who
invented the word "scientist") asserted
that, "the whole fabric of chemistry rests
. . . upon the opposition of acids and
bases." Alkalis had been known since
the dawn of Antiquity, in the form of
soda (sodium carbonate) and potash
(potassium carbonate), which could be
obtained from aqueous extracts of ash
by-products from soap and glass
manufacture. Known as fixed alkalis
(because they were nonvolatile), these
contrasted with volatile alkalis, such as
ammonia, formed from the decomposition of urine. To these were later added
alkaline earths—the name given to
calcium carbonates from chalk and
limestone (and later to salts of magnesium and other metals).

Meanwhile organic acids such as vinegar
and lemon juice had been known since
ancient times; to these the Islamic and
medieval European alchemists had
added spirit of vinegar (purified acetic
acid) and inorganic acids, which were
much more powerful. These included
spirit of salt (hydrochloric acid), spirit
of nitre (nitric acid) and spirit of vitriol
(sulfuric acid). Their reactions with
the alkali could be quite violent, with
effervescence (release of "airs") and
heat, and they fitted nicely with
alchemical doctrines of antagonism
between "male" and "female" principles. Explaining how they worked and
what made them acidic or alkaline
proved difficult because they could only
be defined in circular fashion, until
Robert Boyle (see pp. 74–75) discovered
a way of classifying them using plant
infusions (a forerunner of today's litmus
test). He found that syrup of violets was
blue when neutral but turned red with
acid and green with alkali.

## ACIDS AND BASES IN EVERYDAY LIFE

There are probably plenty of acids and bases, or alkalis, around your home. Typical household acids include vinegar (acetic acid); carbonic acid, found in sodas and carbonated water, formed when bubbles of carbon dioxide dissolve in water; acetylsalicylic acid, aka aspirin; and sulfuric acid, found in car batteries. Typical household alkalis include ammonia, used as a cleaner; lye (sodium hydroxide), another cleaner; baking soda (sodium bicarbonate); and stomach-settling antacids such as calcium carbonate and aluminum hydroxide.

$$HCl(aq) \longrightarrow H^+ + Cl^-$$
$$NaOH(aq) \longrightarrow Na^+ + OH^-$$

In this scheme reactions between acids and bases are neutralization reactions, because the products are water and a neutral salt:

$$HCl(aq) + NaOH(aq) \longrightarrow$$
$$H_2O(l) + NaCl(aq)$$

The water is produced when the hydrogen ion from the acid and the hydroxide ion from the base get together.

Arrhenius's model is accurate for acids and bases that are aqueous solutions, but it is possible to have acid-base reactions between gases; for example, ammonia and hydrogen chloride gases produce the solid ammonium chloride. To account for such reactions a more general theory of acids and bases is needed: known as the Bronsted-Lewis theory, it sees acids as proton donors (i.e. electron acceptors) and bases as proton acceptors (i.e. electron donors). The $H^+$ and $OH^-$ of the Arrhenius model are simply specific examples of a proton donor and an electron donor, respectively.

## Mechanisms of Acidity and Alkalinity

Explanations of how acids and bases work have evolved over time. First came alchemical ideas about the antagonism of opposing principles, followed by theories linking acidity to phlogiston (see pp. 88–89). Oxygen was seen as the arbiter of acidity until John Davy showed that hydrochloric acid (HCl) lacked oxygen, proposing instead that hydrogen caused acidity. In the late 19th century, Svante Arrhenius defined an acid as a substance that yields a hydrogen ion (in other words, a proton, $H^+$), and an alkali as one that dissolves to yield a hydroxide ion ($OH^-$), as shown in the chemical equations for hydrochloric acid and sodium hydroxide:

# 9 Acid-Base Titration

## THE PROBLEM:

Jacqueline works in a quality control lab at a food factory and needs to determine the exact concentration of 50 ml of a hydrochloric acid solution by titration against a standard 0.1000 molar (0.1000 M) solution of alkaline sodium hydroxide. She can use an "indicator" to show how much alkali she needs to neutralize the hydrochloric acid. So, how does she then work out the concentration?

## THE METHOD:

The concentration of an acid or base in solution can be determined by a process of titration. This involves adding a base solution to an acid solution (or vice-versa) and using simple indicators to track the change in either color or pH balance. In this exercise, a base solution (sodium hydroxide) will be added to the hydrochloric acid solution until either the color of the solution changes (this is known as the endpoint) or until the acid is neutralized (this is known as the equivalence point). The endpoint is highlighted by using a chemical indicator such as phenolphthalein,

while the equivalence point can be gauged using a pH meter.

In this exercise, 1 mole of sodium hydroxide (NaOH) will react with 1 mole of hydrochloric acid (HCl) in aqueous solution, forming salt and water. This means that, given an equal concentration of both solutions, 50 ml (for example) of the base will completely neutralize 50 ml of the acid. This is known as a neutralization reaction. Note that in other types of reaction, the equivalence point doesn't mean that the solution will be neutralized. The equation is written as:

$$NaOH(aq) + HCl(aq) \longrightarrow NaCl(aq) + H_2O($$

Jacqueline adds the sodium hydroxide solution to the hydrochloric acid solution using a burette, which allows the precise quantity of solution to be measured. Once the end, or equivalence, point has been reached, she can calculate the exact concentration of the hydrochloric acid solution. To do this, she simply needs to know the concentration of the sodium hydroxide solution that has been added and the volume required to completely neutralize the acid solution.

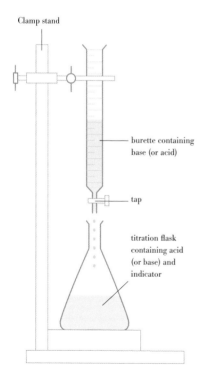

Clamp stand

burette containing base (or acid)

tap

titration flask containing acid (or base) and indicator

## THE SOLUTION:

After titration, Jacqueline establishes that 25 ml of the base solution neutralizes the acid solution. At equivalence, the number of moles of acid is equal to the number of moles of base. Given this 1:1 molar ratio, The equation that Jacqueline needs to solve is:

$$M \text{ NaOH} \times V \text{ NaOH} = M \text{ HCl} \times V \text{ HCl}$$

Here, $M$ is the concentration of the base and acid and $V$ is the volume of the solution.

Jacqueline has established the following: the concentration of the base solution is 0.1 moles; the volume of the acid solution is 50 ml; 25 ml of the base solution is required to neutralize the acid solution. With these values, the equation can be solved:

$$0.1 \times 25 = M\text{HCl} \times 50\text{ml}$$

$$M\text{HCl} = (0.1 \times 25) / 50 = 0.05$$

Therefore, the concentration of the 50 ml sample of hydrochloric acid solution is 0.05 mol dm$^{-3}$.

• Apparatus for titration of acids and bases. The burette makes it possible to determine precisely how much reactant it requires to neutralize the target solution.

# Robert Boyle

The centuries of accumulating alchemical discoveries seemed to be building toward something: a break with the past and a new, scientific approach to chemistry. This paradigm shift was personified in the life and career of Robert Boyle, an Anglo-Irish nobleman whose discoveries in experimental and pneumatic chemistry have led to him being regarded as the "father of scientific chemistry."

## His Greatest Delight

The fourteenth son of the immensely wealthy Earl of Cork, Robert Boyle (1627–1691) was expensively educated and traveled Europe as a youth. Extremely pious throughout his life, he concentrated on theological studies in his early years before coming into contact with a circle of natural philosophers and alchemists who drew inspiration from Francis Bacon (see pp. 61–62). In particular, in the early 1650s, he fell in with the American alchemist George Starkey (1628–1665), who schooled him extensively in the lore and practice of alchemy, teaching him the skills and art of the "chymist."

At this time "chymistry" was seen as a suspect pursuit, mixing the vulgar skills of the artisan (such as the apothecary or metallurgist) with the esoteric quest for the Philosopher's Stone and the transmutation of gold.

In the late 1650s Boyle moved to Oxford, England, and it was here that he started to make his experimental breakthroughs, increasingly articulating his lifelong goal: to marry the experimental expertise and insights of the practical "chymists" with the loftier ambitions of natural philosophy (which aimed to produce "systems of the world"—explanatory models of the universe). In 1668 he moved to London and was a founder member of the Royal Society, the scientific body regarded as the crucible of the Scientific Revolution, which was largely based on Boyle's informal circle of fellow thinkers. Widely published in England and across the continent, he became quite famous and was memorably described in John Aubrey's *Brief Lives* (1681): "He is very tall (about six foot high) and straight, very temperate, and virtuous, and frugal ... His greatest delight is Chymistry. He has ... a noble laboratory, and several servants (Prentices to him) to look to it. He is charitable to ingenious

"'Tis evident, that as common Air when reduce'd to half its wonted extent, obtained near about twice as forcible a Spring as it had before."

—*Robert Boyle*

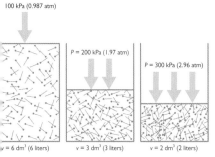

100 kPa (0.987 atm)

$P$ = 200 kPa (1.97 atm)

$P$ = 300 kPa (2.96 atm)

$v$ = 6 dm³ (6 liters)     $v$ = 3 dm³ (3 liters)     $v$ = 2 dm³ (2 liters)

- Demonstration of Boyle's law—to halve the volume requires double the pressure. Tripling the pressure reduces the volume by a third.

men that are in want, and foreign Chymists have had large proof of his bounty, for he will not spare for cost to get any rare Secret."

## The Sceptical Chymist

Boyle's chemical discoveries included a color test for acids, numerous medical treatments, and extensive work with air pumps and vacuums, which led him to formulate what is now known as Boyle's law: "pressures and expansions [are] in reciprocal proportion." What this means is that for a gas there is an inverse relationship between pressure and volume; in other words, if you compress a volume of gas by half, its pressure will double. He also gave one of the first clear definitions of an element: "Certain primitive or simple, or perfectly unmingled bodies; which not being made of any other bodies, or of one another, are the ingredients of which all those called perfectly mixt bodies are immediately compounded, and into which they are ultimately resolved."

Boyle became an ardent supporter of the Atomist philosophy, although he preferred the term "corpuscles" to "atoms." For Boyle, corpuscularianism represented a break with pre-scientific Aristotelian chemistry, and to

advance and defend this new (and ancient) theory he used experiments. In what he called his "Essay on Nitre," for instance, he demonstrated how the chemistry of saltpeter (a component of gunpowder) could be explained entirely in terms of the size and motion of corpuscles, without the need for any of the explanations in terms of "forms" and "qualities" associated with traditional scholastic natural philosophy.

It was this effort to replace the qualitative thinking of the old ways that led to his most famous work *The Sceptical Chymist* (1661), a dialogue attacking the doctrine of four elements and the Paracelsian *tria prima* (see p. 59), while at the same time attempting to persuade "vulgar chymists" that they needed to adopt a more philosophical approach to the study of nature. His reputation as the "father of scientific chemistry" rests less on his actual discoveries than on this philosophical agenda.

### BOYLE'S WISH LIST

Recently displayed at the Royal Society in London were notes by Boyle that constitute a sort of "wish list" for future scientists. Alongside encouragement to develop mind-altering drugs and pain relief, he also recommended research aimed at "Attaining Gigantik Dimensions," believed to be a reference to the possibility of enlarging the human race.

**THE PROBLEM:**

In Exercise #5, we saw that Amy monitored the rate at which hydrogen gas was consumed in her reaction. She now wants to examine the effect of increasing just the pressure on the volume of hydrogen gas. A sample of hydrogen gas collected in a 350 $cm^3$ container exerts a pressure of 103 atmospheres at 600°C. What would be its new volume at 150 atm pressure?

**THE METHOD:**

Amy knows from Boyle's law that the pressure (P) of a gas is inversely proportional to its volume (V) at constant temperature, as expressed in the formula P x V = [a constant]. This can also be expressed as $P_1V_1 = P_2V_2$. Using this equation, she can thus calculate the new volume.

**THE SOLUTION:**

Amy decides to use the formula $P_1V_1 = P_2V_2$, and then lists the values that are given, and those that are unknown. Thus, in this case:

$P_1$ = 103 atm
$V_1$ = 350 $cm^3$
$P_2$ = 150 atm

$V_2$ is the unknown figure she hopes to determine.

   Next, she can predict what should happen. Since the pressure is going up by almost a half, the volume should go down by a bit less than half because pressure is inversely proportional to volume at a constant temperature. Next, she rearranges the original formula to find the unknown $V_2$ in order to then solve this equation for $V_2$, and finally round the value of $V_2$ obtained to the correct number of significant digits.

Thus, $P_1V_1 = P_2V_2$ and therefore $V_2 = P_1V_1 / P_2$

So in this case she can plug in her known values:

$V_2 = 103$ atm x $350$ cm$^3$ / $150$ atm = $240.333333$ cm$^3$

Therefore, the new volume is now $240$ cm$^3$ and Amy can proceed with her work.

In practice, Boyle's law is normally combined with Charles's law and Gay-Lussac's law into the so-called "combined gas laws," represented mathematically as $P_1V_1 / T_1 = P_2V_2 / T_2$, where T is temperature. At a constant temperature in which $T_1$ is the same value as $T_2$, the formula is the same as that of Boyle's law used by Amy in her calculation. Boyle was known as a natural philosopher and although his work touched on what we now refer to as chemistry, physics, and technology, and even theology, it had its roots in the alchemical tradition. Nevertheless, today Boyle is recognized as the first modern chemist and his "law" is among the most important across many applications, from the chemical industry to engineering.

• Demonstration of the combined gas law, showing that the ratio between the temperature of a system and the product of pressure multiplied by volume remains constant.

medium temperature
medium pressure

low temperature
high pressure

high temperature
low pressure

# COVALENT AND IONIC BONDS

Before going any further with the development of scientific chemistry it is probably time to introduce the concept of chemical bonding, describe the two main types of bond, and explain the underlying principle of chemical bonding, which is the tendency for electrons to distribute themselves in space around atoms so as to lower the total energy of the group.

## The Octet Rule

Just as water "likes" to run downhill until it occupies the position with the lowest possible energy given the immediate surrounding landscape, so atoms like to achieve the lowest possible energy configuration when matching up with other atoms to form compounds. The overall energy level of the group of atoms that make up a compound is what determines whether they will react to form chemical bonds with one another in the first place: if the total energy of a group of atoms is lower than the sum of the energies of the component atoms, they then bond together and the energy lowering is the bonding energy.

Outside of atomic chemistry and radioactive phenomena, protons and neutrons do not generally move around, so the energy configuration of an atom is largely determined by the distribution of its electrons. Most importantly for chemical bonding, the outermost shell of electrons orbiting an atom (see p. 27) is known as the valence shell, and its completeness determines bond formation and the reactivity of that atom.

Valence shells follow the octet rule, which states that the most stable, lowest energy configuration is for the shell to have eight electrons. The elements in nature that already have eight electrons in their outermost shells, and are thus "full" or "complete," are the noble gases, including helium, neon and argon. Sometimes also known as the inert gases, these elements are noted for their lack of reactivity; it is very hard to get them to react with other elements because they already have stable, full valence shells. As a general rule in chemical bonding, electrons will try to redistribute, or transfer, themselves to "achieve" the valence shell configuration of the noble gas that is nearest in atomic number. Transferability of electrons is the key to bond formation.

## Ties That Bind

The primary evidence that there are two different types of chemical bond came when chemists started testing solutions for their electrolysis properties (see pp. 138–139). It was found that some compounds, such as table salt, dissolve in water to produce conductive

## IONIC VS. COVALENT COMPOUNDS CHECKLIST

**Ionic compounds**
Electrolytes
Generally solid at room temperature
Higher melting point

**Covalent compounds**
Non-electrolytes
Can be solids, liquids, or gases
Lower melting point

solutions, while others, such as table sugar, do not. Compounds that give conductive solutions are known as electrolytes, while those that do not are non-electrolytes. This was a clue to the existence of ionic and covalent bonds.

In an ionic bond between two atoms, one atom completely transfers one or more electrons to another atom, as each atom seeks to achieve a full outer valence shell (to resemble that of its nearest noble gas). The donating atom is seeking to shed its "incomplete" outermost shell of electrons, so that the "complete" shell beneath becomes the new valence shell. The receiving atom is trying to fill up its outer shell so that it becomes complete. For instance, table salt (NaCl) is made up of sodium atoms ionically bonded to chlorine atoms; when they bond the sodium atom loses one electron to achieve a neon-like arrangement of its electrons, while the

chlorine atom gains an electron to achieve an argon-like arrangement. As a result the atoms become ions—a sodium ion that is positively charged (a cation), and a chlorine ion that is negatively charged (an anion), so that the more accurate chemical formula for table salt is $Na^+Cl^-$. Between positive and negative ions there is an electrostatic attraction, and this is what binds the particles together in an ionic compound.

In a covalent bond the electrons are not completely transferred as in an ionic bond. Instead two atoms share a pair of electrons; the electrons effectively take up a new orbit encompassing both atoms. Again, each atom is seeking to achieve a full or nearest-noble gas outer valence shell. A simple example is bromine; it exists in nature as a diatom ($Br_2$) because it is seeking to obtain a krypton-like valence shell. A single bromine atom has seven electrons in its outermost or valence shell; to become like krypton it needs eight, so two bromine atoms share an electron pair, allowing each to fill its octet and achieve a stable, low energy configuration.

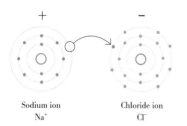

Sodium ion
$Na^+$

Chloride ion
$Cl^-$

• The respective electron configurations of a sodium ion and a chlorine ion, showing how an electron is donated to allow both to achieve a noble gas-like valence shell.

# 11 Atoms, Molecules, and Compounds

## THE PROBLEM:

Buzz is about to take his mid-term science test and has revised the difference between atoms, elements, molecules, and compounds. In the test, he is given the following questions: "What is the difference between atoms, elements, molecules, compounds, and mixtures, and which of the following substances is a molecule or a compound: $H_2$, $O_2$, $CO_2$, and $H_2O$? Is iron(II) sulfide (FeS) a mixture or a compound and what happens when you mix iron with sulfur?"

## THE METHOD:

From Buzz's revision, he remembers that atoms are the smallest particles of matter and that an element is a substance that contains entirely one type of atom. Also, a molecule is formed when two or more atoms are joined chemically and a compound is a molecule that contains at least two different elements. Thus, if the elements that are joined are the same then it is a molecule and if different then it is a compound. Therefore, he deduces that all compounds are molecules but not all molecules are compounds. He also finds that a mixture is a substance made by combining two or more different materials (with no chemical reaction occurring), which can be physically separated into its components, unlike a compound which cannot.

## THE SOLUTION:

Based on this, the answers to the questions are explained. The substances $H_2$ and $O_2$ are both molecules since in $H_2$ two atoms of hydrogen are joined together and, similarly, in $O_2$ two atoms of oxygen are also joined together. In the case of $H_2O$, then this is a compound since two atoms of hydrogen and one atom of oxygen are joined, similarly $CO_2$, with two atoms of oxygen (O) and one atom of carbon (C) joined, is a compound; both are also molecules.

In answer to the other question, the elements iron (Fe) and sulfur (S) can be mixed together into a mixture which can then be physically separated back into iron and sulfur using a magnet.

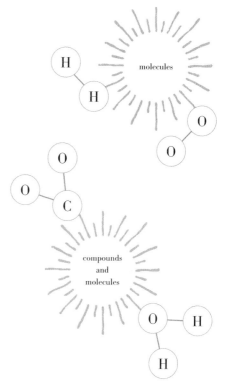

On the other hand, if iron and sulfur are mixed and heated, then the compound iron(II) sulfide (FeS) is formed, which cannot be separated back into iron and sulfur using a magnet, hence FeS is a compound:

$$Fe(s) + S(s) \longrightarrow FeS(s).$$

In nature, another iron sulfide, the compound $FeS_2$, occurs as the mineral pyrite (fool's gold).

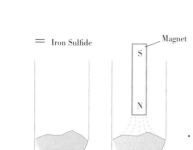

Iron (Fe)  $+$  Sulfur (S)

$=$ Iron Sulfide     Magnet

• It is possible to separate elements of a mixture, such as iron and sulfur, using a magnet. In a compound, the bonds between elements are strong enough to prohibit this.

# THE SCIENTIFIC METHOD

What does it mean to say that chemistry before Boyle was "un-" or "proto-scientific," but that during the 17th century it became a science? What was so different about the chemistry of Jabir or Paracelsus and that of Boyle? The answer lies in a new methodology and a new philosophy, brought together to give a unified system of incredible power: the scientific method.

## The Trouble with Alchemy

We have already touched on some of the unscientific characteristics of alchemy, but it is worth explaining them in detail to highlight the contrast with what came after. At its heart, alchemy is based on *a priori* assumptions and rules (namely, statements and propositions held to be true, even though they have not been tested or proved. For example, the assumption that there are four basic elements, or the belief that there exist correspondences between metals and signs of the zodiac). Alchemical procedures, techniques and recipes stress subjective variables, such as the mental and spiritual state of the experimenter, so that it was believed an experiment might fail because the experimenter was not sufficiently pure of spirit, for instance. According to historian of science Michael White, 'it is this concept, more than any other, which distinguishes alchemy from the orthodox chemistry that superseded it.'

Rather than sharing results and the details of experiments, such as techniques and quantities, so that others could examine, critique, and replicate, as in science, alchemists believed in hiding results and techniques through allegory and encryption. In fact alchemical tracts emphasized the opposite of replicability, stressing the individualism of experimenter and experiments. Finally, alchemists resisted attempts to formalize their art or bring knowledge together into coherent systems or theories.

## The Vanity of Speculations

Alchemy was by no means the worst offender amongst the topics studied by natural philosophers. In medicine, astronomy, biology, and physics the prevailing approach was scholasticism, based on a priori assumptions and the importance of authority. From Francis Bacon to Robert Boyle and his younger colleague Isaac Newton (1642–1727), the new breed of natural philosophers sought to pioneer a new way of doing things. Their agenda, in the words of Bacon, was "to recall natural philosophy from the vanity of speculations," instead favoring experiments that observed nature as it actually was.

"We are certainly not to relinquish the evidence of experiments for the sake of dreams and vain fictions of our own devising."

—*Sir Isaac Newton*

Boyle and Newton were key figures in developing a new, scientific method. Simply stated, the scientific method is that observations of nature (perhaps from an experiment) lead to an initial hypothesis to explain the phenomena, which in turn suggests experiments to test the hypothesis. If experimental evidence does not support the hypothesis, it must be modified or discarded. (Boyle, for instance, was ahead of his time in recognizing the value of unsuccessful experiments.) If the results of experiments support the hypothesis, and they can be replicated, it may graduate to the status of a theory. If repeated patterns can be observed, and better still mathematically quantified, the theory may give rise to laws or axioms. If new evidence comes along that does not fit with the theory, the theory must be modified or discarded. According to its supporters, the scientific method is the only sure route to truth.

• One version, much simplified, of the scientific method in action, stressing the need to report the findings of experiments so that others can attempt to replicate and verify them.

## Without Suspicion of Doubt

This new experimental philosophy was articulated most clearly and powerfully by Boyle: "an Excellent Hypothesis [should] enable a skilful Naturalist to foretell future Phaenomena by the Congruity or Incongruity to it; and especially the event of such Experim'ts as are aptly devis'd to examine it"—and after him by Newton. "For what I shall tell concerning them is not an Hypothesis but most rigid consequence, not conjectured by barely inferring tis thus because not otherwise," wrote the philosopher in a landmark paper on optics, "but evinced by the mediation of experiments concluding directly & without suspicion of doubt." Newton reserved particular scorn for what he called "hypotheses" (in this context meaning speculations unsupported by experimental evidence), expressed in his famous lines: "I frame no hypotheses; for whatever is not deduced from the phenomena is to be called an hypothesis; and hypotheses . . . have no place in experimental philosophy."

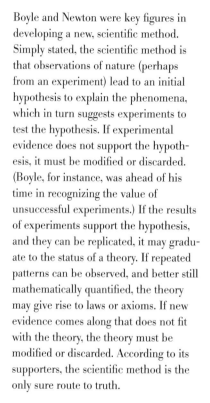

ask a question — do background research — construct a hypothesis — test your hypothesis (experiment) — analyze data / draw conclusion — report your findings

*Chapter*

# Tracking Down the Elements

The Scientific Revolution was the catalyst
chemistry had been waiting for, and the new
science advanced in leaps and bounds. With new
concepts, equipment, and techniques it was
possible to discover new elements at a rate
undreamed of by the ancients. Fortune and glory
awaited those who could reveal new pages of the
Book of Nature, and this chapter relates the
excitement of the race to identify the principles of
the new science.

# Carl Scheele

**In 1669 German alchemist Hennig Brand won fame and fortune by isolating phosphorus from urine. However, it was not until the mid-18th century that a flood of further elemental discoveries was unleashed, as chemical analysis became more sophisticated. Sweden was the initial epicenter of these discoveries, in particular a man who never got the credit he deserved.**

## Kobolds and Kupfernickels

Miners knew a lot about metals and ores, but much of their knowledge was folkloric and superstitious. Two prime examples were *kobolds* and *kupfernickels*—German terms for goblins and gremlins—earth spirits blamed for strange sounds, fumes, and mishaps in the mines. *Kobolds*, for instance, were held responsible for the noxious fumes emitted by a false copper ore, the smelting product of which gives glass a vibrant blue color. In 1735 Swedish chemist Georg Brandt (1694–1768) was able to show that the blue color was due to the presence of a metal, which he named cobalt. A similar spirit known in Saxony as a nickel, or "scamp," was blamed for a false copper ore labeled kupfernickel ("copper scamp"). In 1751 Swedish chemist Axel Cronstedt (1722–1765) analyzed the ore to reveal a previously unknown, hard white metal that he named nickel.

## The Apothecary's Assistant

Carl Wilhelm Scheele (1742–1786) was a remarkable and largely self-taught chemist from Pomerania, usually part of Germany but then in Sweden. Despite being one of 11 children in a poor family and receiving little formal education until he became an apothecary's apprentice in 1756, Scheele became an expert chemist and formidable experimenter. After working all over Sweden he took over a pharmacy in the small town of Köping and remained in this post for the rest of his short life, refusing prestigious academic posts.

The list of Scheele's discoveries is remarkable, spanning the fields of organic and inorganic chemistry. It includes arsenic acid, arsine, barium oxide, benzoic acid, calcium tungstate, chlorine, citric acid, copper arsenite (known as Scheele's green), glycerol, hydrogen cyanide and hydrocyanic acid, hydrogen fluoride, hydrogen sulfide, lactic acid, malic acid, manganese and manganates, molybdic acid, nitrogen, oxalic acid, oxygen, permanganates, silicon tetrafluoride, tartaric acid, and uric acid. He also discovered that the action of light modifies certain silver salts (50 years before they were first used in photographic emulsions) and isolated phosphorus from bone ash.

• Swedish apothecary and chemist Carl Wilhelm Scheele, who, despite his remarkable range of achievements and discoveries, toiled in relative obscurity throughout his career.

Scheele was a practical man, not a theorizer or systematizer, and he may not have recognized or definitively identified some of his discoveries, yet it is extraordinary that he received such little credit at the time. Partly this is down to his relative obscurity as an apothecary in a small town in Sweden, writing in German, but largely it was due to a stroke of remarkable ill fortune. Finally committing his work to paper in 1773 in his book *A Chemical Treatise on Air and Fire*, Scheele had to wait four long years for it to be published as he waited for his mentor, Swedish chemistry professor Torbern Bergman (1735–1785), to finish writing the preface. Unfortunately this meant that Scheele's isolation of oxygen, though predating Priestley's, was only announced three years after the Englishman had published his own paper (see p. 102), and it is Priestley who has since been credited with the discovery.

## Empyreal Air

Nonetheless Scheele had made this discovery prior to 1771, separating air into two main components, one of which he found strongly supported combustion, producing "so vivid a light that it dazzles the eyes."

"Since this air is absolutely necessary for the generation of fire, and makes about one-third of our common air, I shall henceforth, for shortness sake call it empyreal air [fire air] . . ."

—*Carl Scheele*

### THE SECRET FIRE

Taking his cue from the Paracelsian "doctrine of signatures," Hennig Brand (ca. 1630–1692) looked to urine as a potential source for the Philosopher's Stone. He let 60 tubs of urine putrefy in his basement then boiled them to produce a paste, which he heated and drew into water to condense the vapors, arriving at a waxy white substance that glowed in the dark. He named it phosphorus, after the Greek for "bringer of light."

He went on to produce this empyreal air in a series of experiments over the years 1771–2, including heating saltpeter, heavy metal nitrates, and mercuric oxide, and showed that it played a role in plant and fish respiration.

Like Priestley, Scheele failed to grasp fully the implications of his finding, advocating the phlogiston theory of combustion and reduction (see pp. 88–89). When he isolated chlorine in 1774 he mistakenly thought it was a dephlogisticated compound of empyreal air; the gas was not recognized as an element until Davy's work on hydrochloric acid (see p. 71). Eventually, Scheele's hands-on method of analysis, which involved tasting many of his discoveries (including hydrocyanic acid, one of the deadliest poisons known), caught up with him. According to the *Cambridge Dictionary of Scientists*, "His death at 43 is unsurprising; he was a fanatical, prolific, and probably unwise chemical discoverer."

# PHLOGISTON

Phlogiston was a hypothetical substance believed to be the principle agent involved in combustion, reduction and respiration. Although now widely derided as a scientific dead end that led major figures in chemistry astray, in many ways this landmark in the history of chemistry was a valuable hypothesis and an instructive illustration of the power of the scientific method.

## Ash and Calx

The processes of combustion, rusting, respiration, fermentation, and calcination (heating a substance to high temperature but still below its melting point) fascinated alchemists and early chemists alike. These phenomena were clearly interlinked, and uncovering the hidden relationships between them—and in particular the nature of a common principle—promised to reveal one of the most profound secrets of nature. Now we know that the common principle is oxygen, and that, for instance, combustion of wood is a form of oxidation in which carbon is oxidized to gaseous carbon dioxide, leaving only ash behind, while what early chemists called a calx—the powdery substance left behind when a pure metal is heated in air—is a metal oxide. But before oxygen had been discovered, alternative hypotheses were just as reasonable.

Among the first to propose such a hypothesis was Joachim Becher (1635–1681), who in 1667 argued that combustible substances contained an active principle, which he labeled terra pinguis ("fatty earth"). Van Helmont adopted this hypothesis and coined the term "phlogiston," from the Greek for "inflammable," but it was German physician and chemist Georg Ernst Stahl (1660–1734) who first expounded a proper theory of phlogiston. Stahl observed the process of combustion and saw that burning a piece of charcoal produces flames and smoke, and leaves behind a quantity of ash far smaller than the original matter. Clearly, he argued, the heat and smoke represented something being driven off, and that something was the principle of combustibility—phlogiston. Charcoal, he deduced, is effectively composed of ash and phlogiston. Reduction of the products of combustion—as when heating a calx with charcoal to produce pure metal—is a reversal of the process, with the calx absorbing phlogiston from the charcoal to produce the metal (suggesting, by extension, that the calx was the pure, elemental form, while the metal = calx + phlogiston).

## DELUSION OR TRIUMPH?

Phlogiston is often derided as a delusion and snare, or touted as proof that respected scientists can be turned into fools and charlatans by a pseudo-scientific theory. In practice, however, the phlogiston affair can be seen as a triumph for the scientific method. Despite entrenched support it was possible to overturn and discard a popular, attractive and apparently logical hypothesis, simply through a process of trial by experiment. Once evidence had accumulated sufficiently and landmark experiments had conclusively demonstrated its falsehood, the scientific community moved on from phlogiston.

## Phlogiston Falls Down

As the first rational explanation for combustion and calcination, phlogiston made sense. It fulfilled the criteria of a proper scientific theory—it was coherent, fit the observable facts, and suggested testable predictions. French chemical encyclopedist Pierre-Joseph Macquer believed it had "changed the face of chemistry" and urged chemists to search for new substances that would prove its existence. Joseph Priestley adapted the theory to explain the gases he had discovered (see p. 103). But even during Stahl's lifetime doubts were raised, and as chemistry became more precise, with more careful weighing of products and reactants, the theory ran into trouble.

Stahl's theory suggested that combustion should result in loss of mass as phlogiston is driven off, while reduction of calx with charcoal should add mass as phlogiston is absorbed, but experimental results showed the opposite. In 1763 the secretary of the Paris Academy of Sciences called this one of the "true paradoxes of chemistry." (We now know that metal heated in air gains mass because it has combined with oxygen, whereas reduced calx loses mass because it loses oxygen.) Stahl and phlogiston's advocates tried to argue around this, suggesting that phlogiston might have zero or negative mass, and simply be a massless principle akin to "caloric" (a contemporary term for heat energy). Phlogiston was finally killed off in 1774 by Lavoisier's oxygen theory (see p. 104).

Political Portraiture  N°4.

DOCTOR PHLOGISTON,
The PRIESTLEY politician or the Political Priest.

• Late 18th-century cartoon satirizing Joseph Priestley as "Doctor Phlogiston, The Priestley politician or the Political Priest," trampling on the Bible while radical tracts spill from his pockets.

## 12 We Have Combustion!

### THE PROBLEM:

Sadie had heard of the old theory that a fire-like element called "phlogiston" is found in combustible materials and is released on burning in air. It was disproven when it was found that metals burned in air (oxygen) gain mass rather than lose it. To prove this, two separate 5-gram samples of magnesium powder and aluminum powder, on being burned in air, both produced white metal oxides, weighing 8.33 grams and 9.44 grams respectively. From this, how does Sadie deduce the composition of, and equations for, the oxides of magnesium and aluminum that formed?

### THE METHOD:

Since magnesium oxide and aluminum oxide are the only products formed, we can find the increase in weight after the two powders are burned. From the law of conservation of mass, which states that in a chemical reaction the mass of the products equals the mass of the reactants, then 5 grams of both metals have reacted with $x$ or $y$ grams of oxygen, where $x$ and $y$ are the respective increases in mass after combustion of magnesium and aluminum metal.

Knowing the "mole ratio" of metal and oxygen combined, we can obtain the formula.

### THE SOLUTION:

First, we need to find the mass of oxygen in both the oxides. Given that the extra mass acquired after the metals are burned is in fact oxygen, this is simple: we simply note the difference in weight before, and after, burning.

In the first case, 5 grams of magnesium converted to 8.33 grams of magnesium oxide after it was burned. So the mass of oxygen is 3.33 grams

The next step is to find the chemical formula. For this we require the mole ratio, which can be found by first finding the number of moles in 5 grams of magnesium and 3.33 grams of oxygen.

The number of moles in a quantity of a particular element is calculated by dividing the weight (in this case, the 5 grams of magnesium) by the atomic mass of a single atom of that element (see p. 121 for more details).

The atomic masses for the elements in question are:

magnesium = 24
aluminum = 27
oxygen = 16.

The number of moles in 5 grams of magnesium is:

$$5 / 24 = 0.208 \text{ moles}$$

And the number of moles in 3.33 grams of oxygen:

$$3.3 / 16 = 0.206$$

As you can see, the number of moles is almost identical for both elements, so we can say with a good degree of accuracy that the ratio is 1:1. Given this parity, the formula is MgO.

In the second case, 5 grams of aluminum converted to 9.44 grams of aluminum oxide, a difference of 4.44. Therefore the mass of oxygen in this reaction is 4.44 grams. Using the same procedure as before, the ratio of aluminum (Al) to oxygen (O) moles combined is:

$$(5 / 27):(4.4 / 16) = 0.185:0.275$$

0.185 into 0.275 goes 0.67 times–a ratio of 2 : 3. The formula is therefore $Al_2O_3$.

The combustion of magnesium and aluminum in air is thus:

$$Mg(s) + \tfrac{1}{2} \, O_2(g) \longrightarrow MgO(s)$$
$$\text{or}$$
$$2 \, Mg(s) + O_2(g) \longrightarrow 2 \, MgO(s),$$

$$2 \, Al(s) + 3 \, O_2(g) \longrightarrow Al_2O_3(s)$$

• The chief commercial use of magnesium today is in the production of aluminum-magnesium alloys, which are valued for their lightness and strength.

# CARBON DIOXIDE

Pneumatic chemistry was gathering pace as increasingly sensitive instruments and sophisticated techniques allowed the analysis of gases (or "airs" as they were known). Whereas alchemists like Van Helmont had been able to surmise the existence of different types of airs in qualitative fashion, the new breed of scientists were able to prove their existence quantitatively.

## Fixed Air

In 1754–56 the Scottish chemist Joseph Black (see box) presented an impressive series of experiments in which he obtained an unknown gas from heating chalk (calcium carbonate) to produce quicklime (calcium oxide). Because it was contained within the solid until released by heating, he called the new gas "fixed air"; it was the same gas described as spiritus sylvestre by Van Helmont a century earlier—what we would now call carbon dioxide ($CO_2$).

In his 1756 paper "Experiments upon Magnesia Alba, Quicklime, and Some Other Alkaline Substances," Black showed a complete cycle of chemical transformations of this new air. Having

---

• **MEET JOE BLACK**

*Born to a Scottish-Irish wine merchant in Bordeaux, France, Black (1728–1799) trained as a physician and progressed through chairs in anatomy, medicine, and chemistry at Glasgow and Edinburgh Universities. His work on causticization and the isolation of fixed air formed his doctoral dissertation, and made him the founder of pneumatic chemistry according to some* accounts *(although Van Helmont and others are also awarded this accolade). After around 1760 he devoted most of his research to physics, leading to his breakthroughs in the field of heat (see pp. 98–99). He was a popular lecturer and his collected talks were published in 1803.*

• Joseph Black proved that carbon dioxide was involved in the processes of life—including breathing, photosynthesis, and fermentation.

decomposed chalk to produce quicklime he then showed that he could reverse the process and recombine the fixed air with quicklime to produce chalk. He further demonstrated that the same fixed air, which he identified by weighing it (a feat requiring ultrasensitive scales), was the product of combustion, fermentation, and respiration. Although he did not pursue research into fixed air, he correctly surmised that it formed a component of the atmosphere ($CO_2$ makes up around 0.04% of fresh air by volume).

Black's experiments with fixed air were part of his research into "causticization"—the opposite of acidification. He described how carbonates (which he defined as mild alkalis) are causticized (made more strongly alkaline) when they lose fixed air (carbon dioxide), whereas when they take up fixed air they are reconverted into mild alkalis. Black also demonstrated how release of carbon dioxide was the cause of effervescence when limestone was added to acids.

## Breath of Life

In a dramatic demonstration of his claim that fixed air was the product of respiration, Black exhaled into a jar of limewater (a solution of slaked lime, aqueous calcium hydroxide), which turned cloudy as tiny particles of chalk formed within. This is still the standard test for presence of carbon dioxide:

$$CO_2(g) + Ca(OH)_2(aq) \longrightarrow CaCO_3(s) + H_2O(l)$$

If you continue to bubble carbon dioxide through the mixture it will turn clear again, as calcium carbonate and carbon dioxide react together to form calcium hydrogen carbonate, which is a colorless solution:

$$CO_2(g) + CaCO_3(s) + H_2O(l) \longrightarrow Ca(HCO_3)_2(aq)$$

Hardness of water is caused by similar reactions, where rainwater is naturally acidified by carbon dioxide and then reacts with limestone in the ground.

# Henry Cavendish

**The man who made the greatest strides in the field of pneumatic chemistry was an eccentric English millionaire, generally regarded as the greatest man of science since Newton but pathologically shy and retiring. Henry Cavendish unlocked the secrets of the atmosphere, created water, and weighed the Earth, but could not bring himself to speak to or even look at a woman.**

## Airs and Graces

Black's elegant experiments inspired English aristocrat Henry Cavendish (1731–1810) to begin his own research into airs. Cavendish was the grandson of the Dukes of Devonshire and Kent and destined to become one of the richest men in Britain when his father died. His only interests, however, were scientific, and he was a noted eccentric and recluse (see box). Using the pneumatic trough, a device invented by Stephen Hales

(see pp. 64–65) which captured gas in an upturned vessel over a water tank, Cavendish studied what he called "factitious airs," including Black's fixed air, produced by reacting acids and bases, and "inflammable air," produced by mixing metal with acids. By measuring the amount of water displaced, Cavendish was able to calculate the specific gravity of the airs (their density relative to the atmosphere as a whole), showing that his inflammable air was the lightest substance yet discovered.

We now know that Cavendish's inflammable air was hydrogen, but he subscribed to the phlogiston theory and

• The pneumatic trough, shown here in an illustration from 1727, permitted Cavendish to measure the properties of gases with much greater accuracy.

## • WEIGHING THE EARTH

Cavendish is celebrated as much for his achievements in physics as chemistry. He made important discoveries in electricity but never published them, and he worked to advance the Newtonian view of energy and atoms, identifying heat with vibration of particles. One of his last experiments involved a delicate apparatus with large spheres attached to a beam, which allowed him to observe the gravitational interaction between objects and calculate the gravitational constant. This in turn made it possible to calculate the mass of Earth.

## THE SHY SCIENTIST

Cavendish cut an odd figure in society. A fellow of the Royal Society described him: "His dress was of the oldest fashion, a grayish green coat, a small cocked hat, and his hair dressed like a wig (which possibly it was) with a thick clubbed tail. He never appeared in London unless lying back in his carriage. He probably uttered fewer words than any other man." Cavendish was so shy that when he forced himself to attend Society meetings he would issue "shrill cries" as he entered a room full of people. If looked at directly he would "retire in great haste," and if spoken to he would flee to the safety of home. He was especially shy around women, communicating with his housekeeper only through notes and forbidding female servants to approach him. When he bumped into a maid on the stairs one day he ordered construction of a back stair to the house.

His immense wealth was of no consequence to him; in one famous anecdote his bankers felt they should approach him on the subject of his balance—the largest in the country—and inquire whether he had any instructions for its deployment. Irritated at being bothered, his reply was that if they found the balance an inconvenience he could remove it elsewhere. Even his death was eccentric; knowing his end was near he gave strict instructions to be left alone until a time when he had calculated he would be dead. When a concerned butler came in half an hour early he was sent away with a flea in his ear.

believed that he might have discovered phlogiston itself. Later he identified the primary components of the atmosphere (aka "common air"), oxygen and nitrogen, although he called them dephlogisticated and phlogisticated air respectively. In 1781 he combined inflammable air with common air and lit the mix with an electric spark, producing water droplets. Measuring the remaining gas he showed that about a fifth of the common air had vanished.

Later he repeated the experiment with just inflammable and dephlogisticated air (hydrogen and oxygen), producing pure water, and thus finally disproving the ancient concept that water was one of the basic elements (although he wrongly maintained that it was the combination of phlogiston and dephlogisticated air that produced water). He even noted that a tiny remnant of common air was inert; a century later this was identified as argon.

"He was acute, sagacious and profound, and I think the most accomplished British philosopher of his time."—*Sir Humphry Davy*

# WATER

The most familiar liquid is also one of the most unusual, in terms of its chemistry and physical properties, and as a result is the most important substance on Earth. Water occupies a central role in chemistry; known as the universal solvent, it mediates phenomena such as acidity and alkalinity. For life on Earth it is essential in a host of different ways.

## Angular, Polar, and Sticky

The type and distribution of bonds in a compound determine its structure and shape, which in turn determine the properties of the compound. Water, which is composed of an oxygen atom covalently bonded to two hydrogen atoms, has an angular shape:

$$O \diagup^{H}_{\diagdown H}$$

Rather than a linear shape: H−O−H. (In fact the angle of the bonds in the gas phase is 105°.) Its unusual properties, so vital to life on Earth, are a direct result of this angular shape, which effectively gives the molecule sides or "ends." Because oxygen is more electronegative than hydrogen it more strongly attracts the electron pairs in each covalent bond, so that they are pulled closer to the oxygen atom. This in turn gives the oxygen a partial negative charge, while the hydrogen atoms each have a partial positive charge—the molecule as a whole thus has a negative pole and a positive pole, making it a dipole.

One consequence of this polarity is that water molecules interact with each other, with the partially positively charged hydrogen atom of one molecule attracted to the partially negatively charged oxygen atom of another. This interaction is called a hydrogen bond.

Hydrogen bonding between water molecules accounts for many of the unique properties of water, such as its unusually high boiling point. Generally the boiling point of a liquid relates to its molecular weight (see pp. 126–127), and substances with equivalent molecular weight to water are usually gases at room temperature. Water, by contrast, is liquid across a very wide range of temperatures, making it a stable medium for life on Earth. Hydrogen bonding also gives water a very high heat capacity (the amount of heat needed to change its temperature) and a high heat of vaporization (the amount of heat energy needed to make it change phase; see pp. 98–99 for more on heat). As a result, bodies of water can absorb large amounts of heat and release it slowly; on a global scale this helps to prevent the kind of wild swings in day and night temperature seen on other planets, and moderate longer-term climate variation. When water freezes the hydrogen bonds lock the molecules into a rigid array that is less dense than the liquid form, which means that ice floats on water instead of sinking. As a result only the top of a body of water will freeze, while the rest is insulated.

## Universal Solvent

Polarity and its angular shape make water a powerful solvent for both ionic and polar covalent substances. Its partially charged poles allow it to interact strongly with ions; when an ionic substance dissolves in water, water molecules surround the anions and cations with their positive and negative poles respectively. A similar process allows water to dissolve polar covalent substances; many organic compounds, such as sugars, alcohols, and proteins, contain O-H and N-H bonds, which are polar, so they will dissolve in water. An ion surrounded by water molecules is known as a hydrate; it is more correct to describe an ion such as $Cu^{2+}$ (ionized copper) as $[Cu(H_2O)_6]^{2+}$. Many

## LIGHT TOUCH

Water strongly absorbs infrared light but is transparent to visible and near ultraviolet radiation. Water vapor in the air thus lets through solar radiation during the day to heat the planet while restricting heat loss at night, again acting to maintain relatively even temperatures across the day-night cycle. Water vapor is also a greenhouse gas (see p.93).

inorganic substances can form crystalline hydrate solids.

Liquid water also dissociates to a small degree, so that $H_2O \rightleftharpoons H^+ + OH^-$. (The $\rightleftharpoons$ symbol shows the reaction is reversible and takes place in both directions, so that there is an equilibrium between the two states.) Substances that increase the number of $H^+$ ions are acidic and those that increase the number of $OH^-$ ions are basic or alkaline.

## • JESUS LIZARDS FEEL THE TENSION

*Water molecules in the liquid phase are more strongly interlinked and interactive than most other liquids, and this is most noticeable at the surface of a body of liquid water. Because the molecules at the surface are only attracted downward and sideways, whereas all the other molecules are attracted in all directions, water has a very high surface tension. Not only does this allow small creatures (such as Jesus lizards) to walk on the surface of water, it also means that evaporation rates for water are much lower than would be expected, which helps to keep much of Earth's water in the oceans rather than in the atmosphere.*

# HEAT

For the ancients, fire had been one of the primal elements, and although by the mid-18th century the Classical elements had been superseded, chemists still puzzled over the nature and behavior of heat. Intriguing observations and clever experiments would shortly reveal important aspects of the chemistry of heat. Indeed the principles of heat are fundamental to chemistry as a whole.

## The Scientist in the Brewery

Thermometer pioneer Daniel Gabriel Fahrenheit (1686–1736) had made a curious observation. He noted that the temperature of supercooled water, which immediately changed phase to ice when shaken, shot up to 32° on his scale. Joseph Black (see pp. 92–93) followed up this finding with research of his own, which he conducted in a brewery to be sure of steady warmth. Black also observed an apparent mismatch between the heating of water (specifically ice) and its temperature. When he melted ice he noted that although it absorbed heat its temperature did not change; in other words, it went from ice at 32°F (0°C to water at 32°F. It seemed that the heat had somehow combined with the water particles in such a way that it was "hidden" from the thermometer, and so he termed it latent heat. In 1761 he was even able to measure this latent heat (known as the latent heat of fusion in the case of a solid-to-liquid or liquid-to-solid phase change), and the following year he measured the latent heat of vaporization (the latent heat involved in the phase change from liquid to gas).

• **THE DIFFERENCE BETWEEN TEMPERATURE AND HEAT**

*Temperature is the average kinetic energy of individual particles in a body or closed system (for example, a bag of gas). Heat is the total amount of energy in the body or system. For instance, a cup of water and a bathtub full of water may be the same temperature but the bath contains much more heat than the cup.*

## THE MISSING MR. CENTIGRADE

Fahrenheit's was the first widely adopted temperature scale; he set the 0° point at the lowest temperature he could achieve, using a mixture of salt and ice. In 1742 Swedish scientist Anders Celsius (1701–1744) proposed that scientific measurements of temperature should be made on a fixed scale based on two naturally occurring points—the freezing and boiling of water (at sea level). Celsius actually suggested 0° as the temperature at which water boils, and 100° as that at which water freezes, but his pupil Martin Strömer inverted the scale and it was adopted across Europe as the Celsius scale. In Britain this scale has been known as centigrade (literally "100 gradations"). Scientists often prefer to use the Kelvin scale, which starts at absolute zero, a theoretical state in which no energy at all is present. On the Kelvin scale (measured in kelvins [K] named for William Thomson, Lord Kelvin [1824–1907]), water freezes at 273 K. 1 K = 1°C = 1.8°F.

Black followed up these discoveries by finding that equal masses of different substances require different quantities of heat to change their temperatures by the same amount. This is now known as the concept of specific heat. The specific heat of a substance is defined as the energy required to raise 1 gram by 1°C.

A more general concept is heat capacity, which is the amount of heat required to change the temperature of a body or substance (specific heat is known as a derived quantity of heat capacity). Energy is measured in calories or joules; so specific heat is measured in calories or joules per g°C. The specific heat of water is 1 calorie/g°C or 4.186 joule/g°C, which is five times higher than aluminum and roughly 10 times higher than iron or copper.

## The Mystery of Heat

Black's discovery of latent heat fit well with contemporary notions of the nature of heat. If it somehow combined with the water particles and was locked away, was it not similar to phlogiston or fixed air? This was how it seemed to 18th-century scientists, who imagined heat as a form of matter (the "matter of fire"), either particles or an elastic fluid. Lavoisier (see pp. 104–105) later formalized this conception of heat by calling it "caloric." Historian of science J. L. Heilbron points out that the new word "only lightly veiled the descent of [this] principle from earlier ideas about the nature of fire." In other words, caloric was the Classical element of fire by a different name. It took another 70 years until chemists came fully to accept that heat was a form of energy, distinct from matter.

# Exercise 13 Fixed Air

### THE PROBLEM:

In addition to work on "latent heat" and "specific heat," Joseph Black identified a gas that he called "fixed air," which was actually carbon dioxide ($CO_2$). In the lab, Tom was curious about how this gas forms and used two methods to prepare it from other substances: (a) By heating calcium carbonate ($CaCO_3$) to red heat; (b) By adding dilute hydrochloric acid to solid calcium carbonate. Starting with 5 grams of $CaCO_3$, how much $CO_2$ would be produced, by mass and volume at 0°C, in each case?

### THE METHOD:

We first need to write out the two equations for (a), the thermal decomposition of $CaCO_3$, and (b), the action of dilute hydrochloric acid (HCl) on $CaCO_3$, to see how much $CO_2$ gas is produced in each case. Then, knowing the molar amounts, we can calculate the amount of $CO_2$ produced.

### THE SOLUTION:

In the equations for (a) the thermal decomposition of calcium carbonate and (b), the action of hydrochloric acid on calcium carbonate, 1 mole of $CO_2$ gas is produced per 1 mole of $CaCO_3$ used:

(a)  $CaCO_3(s) \longrightarrow CaO(s) + CO_2(g)$

(b)  $CaCO_3(s) + 2\ HCl(aq) \longrightarrow CaCl_2(aq) + H_2O(l) + CO_2$

The first step is to calculate the molecular mass of calcium carbonate and carbon dioxide. This is done using the respective atomic weights: Ca = 40, C = 12, and O = 16.

$$CaCO_3: 40 + 12 + (16 \times 3) = 100 \text{ grams}$$

$$CO_2: 12 + (16 \times 2) = 44 \text{ grams}$$

In the 19th century, Amadeo Avogadro established that 1 mole of every gas occupies the same volume, at the same temperature and pressure (see pp. 128–129). It is also known that 1 mole of a gas occupies 22.4 liters at 0°C and 1 atmosphere pressure.

Using this information, we can calculate the volume of $CO_2$ produced in the two equations. At 0°C and 1 atm, 100 grams of $CaCO_3$ will produce 44 grams of $CO_2$ gas with a volume of 22.4 liters. Therefore, 5 grams of $CaCO_3$ will produce:

$$44 / 20 = 2.2 \text{ grams of } CO_2$$

Which occupies a volume of:

$$22.4 / 20 = 1.12 \text{ liters at 0°C and 1 atm}$$

Carbonated (soda) water is made by passing pressurized $CO_2$ gas through water. The pressure increases the solubility and allows more $CO_2$ gas to dissolve than would be possible under standard atmospheric pressure. When a bottle of soda water is opened, the pressure is released allowing the gas to come out of solution creating the characteristic bubbles and fizziness.

• The bubbles in carbonated drinks have remained popular to the present day.

# Joseph Priestley

**In the late 18th century chemistry started to become extremely popular, its discoveries matters of public interest, its rivalries matters of national interest. The man partly responsible for this, Christian minister and political radical Joseph Priestley, inventor of soda water and discoverer of oxygen, would eventually fall foul of the increasing attention that chemists attracted.**

## Promiscuous Airs

The discoveries of Black and Cavendish heralded exciting times for pneumatic chemistry, or as J. L. Heilbron puts it, "Novel airs then began to rise promiscuously." No one was more promiscuous with his discoveries than Joseph Priestley (1733–1804), said to have isolated no fewer than eight new gases. Priestley came from a religious but dissenting background (dissenters were religious radicals who "dissented" from mainstream Anglicanism, and often had radical political ideas as well, with a particular focus on

extending education to the working classes) and worked as a minister and teacher.

On a trip to London in 1766 he met the American scientist Benjamin Franklin and was encouraged to pursue scientific research, initially into electrical phenomena. Shortly after this he got a job as a minister in Leeds, where his house was next to a brewery. Brewers and chemists alike knew that above the surface of their vats of fermenting beer was a layer of special air that bubbled up during the fermentation process. Priestley showed that this air was identical with Joseph Black's "fixed air," carbon dioxide, and with the plentiful supply at his disposal it occurred to him to attempt to artificially simulate the natural effervescence of some mineral waters. Dissolving the carbon dioxide under pressure in water, he succeeded in creating carbonated water. This discovery, which he refused to patent and disseminated widely in his writings and lectures, sparked a European craze for "soda water."

## Gas Man

In 1773 Lord Shelburne offered Priestley a position that would afford him plenty of time for research. Priestly delved further into pneumatic chemistry, perfecting Hale's pneumatic trough and cleverly using mercury in place of water for collecting gases that were water soluble. He also acquired a magnifying glass 1 foot (30 cm) across, which he used to focus the rays of the Sun and thereby generate very high temperatures. Using this apparatus he isolated gases including nitrogen monoxide (NO), dinitrogen monoxide ($N_2O$), aka laughing gas, sulfur dioxide ($SO_2$), and ammonia ($NH_3$).

By 1772 Priestley had made arguably the first observation of photosynthesis, showing that plants produced an "air" that animals required for respiration. In 1774 he was able to synthesize this air by using his glass to heat red calx of mercury, a powder produced by burning mercury in air (today known as mercury[II] oxide, HgO). By making the calx hot enough he was able to make it turn back into mercury; collecting the gas given off he found that it was colorless and odorless but caused a flame to burn very brightly. Further testing revealed that it was "superior" to common air: "I procured a mouse, and put it into a glass vessel, containing two ounce-measures of the air … Had it been common air, a full-grown mouse, as this was, would have lived in it about a quarter of an hour. In this air, however, my mouse lived a full hour … and appeared not to have received any harm from the experiment."

Priestley was conservative in his approach to theorizing, and a committed phlogistonist. He identified his new gas as dephlogisticated air, but

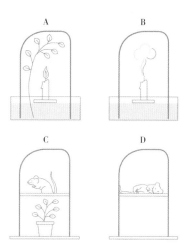

• Illustrations demonstrating the basic principles of photosynthesis. In this theoretical model, both the flame and the mouse rely on the plant to produce oxygen. In turn, the plant relies upon the carbon dioxide that they produce.

on a trip to Paris in 1774 he told Lavoisier (see the next page) about his findings, which led in turn to the identification of Priestley's "superior air" as oxygen, and the dismantling of phlogiston doctrine.

## MOB RULE

Aspects of science became highly politicized in the 18th century, and few scientists were more political than Priestley. As an outspoken and well-known opponent of the established church and supporter of the French Revolution he proved to be a lightning rod for anti-Revolutionary feeling in England. On July 14th, 1791, the second anniversary of the storming of the Bastille, rioters in Birmingham attacked Priestley's home and burnt it down. He fled to London with his family, and was eventually forced into exile, moving to America where he ended his days in comparative isolation, his insistence on clinging to the phlogiston doctrine putting him at odds with scientists in the chemistry mainstream.

# LAVOISIER AND THE CHEMICAL REVOLUTION

The greatest chemist of his age—perhaps ever—discovered no new elements, yet did more than anyone to finally transform chemistry into a science. Known as the father of the "chemical revolution," Antoine-Laurent Lavoisier worked out the role of oxygen, defined the term "element" and introduced scientific nomenclature, before his life was cut short by the Terror of Revolutionary France.

## The Finest Science Money Can Buy

Antoine-Laurent Lavoisier (1743–1794) was the son of a rich lawyer, expensively educated and trained in law before taking up science, initially as a geologist and mineralogist. This led him to chemistry and he set up a laboratory with the aim of gaining admission to the prestigious Academy of Sciences. Admission was duly gained and in 1768 Lavoisier fatefully joined the Ferme Générale, a private consortium that collected taxes for the crown, in order to secure a private income that would fund his researches. Science was becoming increasingly specialist, and consequently expensive—especially the precision instruments upon which Lavoisier's achievements relied.

In 1772 he turned his attention to pneumatic chemistry, experimenting with combustion of phosphorus and sulfur, which, he discovered, gained weight when burned in air. He also found that when he heated litharge (lead[II] oxide, an ore of lead), with charcoal (carbon), it reduced to lead, released gas, and lost weight. Lavoisier called this "one of the most

### • LAVOISIER'S BIGGEST MISTAKE

*Lavoisier was not infallible. A central element in his new system was a hypothetical principle of heat, which he called caloric. Although a type of "imponderable"—an undetectable substance without mass—caloric was supposed to act like a liquid or gas, and Lavoisier claimed that oxygen gas was actually a compound of oxygen and caloric, the latter substance accounting for its phase. This was as much a dead end as phlogiston, and in many ways simply a renamed version of elemental fire (see Heat, pp. 98–99).*

interesting [discoveries] . . . since the time of Stahl," although similar findings had been made 20 years earlier. This finding was at odds with the doctrine of phlogiston, which claimed that the reduction of ore to lead involved addition, not subtraction, and it set Lavoisier on the path to destroying the myth of phlogiston.

## Eminently Respirable Air

In 1774 Lavoisier learned from Priestley of the discovery of "dephlogisticated air." Experimenting with the new gas for himself he soon grasped that here was the common principle underlying the processes of combustion, reduction, respiration, fermentation, and acidity. As Priestley had also done, Lavoisier proved that the new air formed that portion of the atmosphere that supports life, which led him initially to term it "eminently respirable air," and he was able to show

that combustion and respiration both transformed it into the "fixed air" identified by Joseph Black. In 1777 Lavoisier was ready to offer a new "general theory of combustion" to replace phlogiston, along with a new name for his new principle of combustion: "oxygen." His research with the three common mineral acids—nitrous, phosphoric, and vitriolic (sulfuric)—and a newly isolated oxalic acid, from organic sources, had shown that oxygen was present in all four, leading Lavoisier to propose: "I shall henceforth designate dephlogisticated air or eminently respirable air in its state of fixity by the name of acidifying principle, or, if one prefers the same signification in a Greek word, by that of oxygen principle." Oxygen derives from the Greek for "acid-maker."

Armed with this new concept of oxygen, Lavoisier was able to show that the phlogiston doctrine was back to front. Combustion, respiration, and rusting involved the addition of oxygen and reduction its loss. Fixed air was produced by combination of charcoal with oxygen. When he learned how to make water by burning hydrogen in oxygen, the final piece of the puzzle fell into place, and Lavoisier was able to show that water was not simply "de-phlogisticated" air as Cavendish had claimed, but a compound, combining hydrogen (named by Lavoisier from the Greek for "water-maker") with oxygen.

## Elements of Chemistry

The culmination of Lavoisier's chemical career was the publication of his *Traité Elémentaire de Chimie* (*Elements of Chemistry*) in 1789. It set out his findings and his reasoning in clear, logical style, lending irresistible force to his modern, scientific vision of chemistry: "We must trust to nothing but facts: These are presented to us by Nature, and cannot deceive. We ought, in every instance, to submit our reasoning to the test of experiment, and never to search for truth but by the natural road of experiment and observation." Among the important innovations of Lavoisier's chemistry was a new and decisive definition of an element: "the last point which analysis is capable of reaching"—in other words, those substances that could not be broken down any further. He admitted that as technology advanced, some substances previously incapable of decomposition might be shown to be compounds, and indeed several of the substances on his list of 33 elements proved to be oxides. Along the same lines he predicted that several contemporary alkaline earths (basic solids that could not be broken down any further) would prove to be metal oxides, and sure enough Humphry Davy was

• Illustrations, by Madame Lavoisier, of laboratory apparatus from M. Lavoisier's *Traité Elémentaire.*

*Often overlooked in accounts of Lavoisier's exploits is his wife, Marie-Anne Pierrette, née Paulze (1758–1836). When he married her she was just 14, the daughter of a tax farmer in the upper echelons of the Ferme Générale, which he had just joined. Marie promptly embarked on a programme of training and education to prepare for a life assisting her husband in his research; she learned English so that she could read and translate papers and reports from across the Channel, and studied drawing and engraving so that she could record her husband's experiments and illustrate his writings. Over the years she proved her worth both as an able laboratory assistant and as the hostess of a scientific salon. When Lavoisier died she continued her association with science by marrying the British-American physicist Benjamin Thompson, Count Rumford, although the marriage was not a happy one—allegedly she poured boiling water on his flowers—and it ended just a few years later.*

• An illustration by Madame Lavoisier depicting her husband working in their laboratory.

## REVENGE OF THE NERD

Lavoisier was amazingly accomplished. In addition to his chemical research he was diligent in performing his duties as a tax farmer, and also engaged with a range of civic and governmental issues. He perfected the recipe for gunpowder and oversaw improvements in its manufacture, helping to ensure that the young American republic received quality supplies to fight off the British. He was on the committee that investigated and debunked mesmerism, and helped to judge ballooning contests. He helped the Parisian authorities build a wildly unpopular antismuggler wall around the city, which gave rise to a popular saying: *Le mur murant Paris rend Paris murmurant.* ("The wall surrounding Paris is making Paris grumble.") He also helped to devise the metric system. Unfortunately none of these civic accomplishments availed Lavoisier in the face of a vengeful amateur.

Before becoming a revolutionary firebrand, and instigator of the Terror, Jean-Paul Marat (1743–1793) was an aspiring scientific dabbler with pretensions to the Academy of Sciences; his membership had been blocked by Lavoisier. Now a man of terrible power, Marat accused the great chemist of attempting to "imprison" Paris with his unpopular wall. At his trial the judges dismissed Lavoisier's pleas for clemency in order to pursue his scientific research, and his long record of support for, and involvement with, liberal and revolutionary causes counted for nothing when set against his membership of the hated tax-farming operation. Along with his father-in-law he was sentenced to death and executed by guillotine on May 8th, 1794. His colleague Joseph Lagrange famously commented, "It took only a moment to cut off that head, yet a hundred years may not give us another like it."

later able to use the new technology of electrolysis to isolate the alkaline earth metals from their molten salts (see pp. 142–143).

Another great contribution to chemistry as a science was Lavoisier's "balance sheet" approach. Using his expensive and highly sensitive instruments, he perfected the art of measuring the quantities of both reactants and products, whether solid, gas, or liquid, and stressed the importance of quantifying exactly what went into and came out of a reaction. This led him to articulate the principle of conservation of matter:

"We may lay it down as an incontestable axiom, that, in all the operations of art and nature, nothing is created; an equal quantity of matter exists both before and after the experiment; the quality and quantity of the elements remain precisely the same; and nothing takes place beyond changes and modifications in the combination of these elements. Upon this principle the whole art of performing chemical experiments depends: We must always suppose an exact equality between the elements of the body examined and those of the products of its analysis."

### THE PROBLEM:

Oxygen was discovered independently by Scheele and Priestley but later recognized as a chemical element by Lavoisier. It is produced on an industrial scale for chemical manufacturers, hospitals, and other users by the fractional distillation of liquefied air, which separates it from nitrogen and argon. Michelle works in a manufacturing facility and wants to calculate the volume of oxygen gas obtained at 20°C and 1 atmosphere pressure from evaporation of 70 liters of liquid oxygen.

### THE METHOD:

In order to solve the problem, Michelle converts the 70 liters of liquid oxygen into a mass, using its density as a guide. She then finds the number of moles in the 70 liters from its relative molecular mass of $O_2$–32 (at standard temperature and pressure oxygen atoms bond in pairs to form dioxygen). She then calculates the volume of $O_2$ gas at 20°C and 1 atmosphere pressure using the ideal gas law, $PV = nRT$.

### THE SOLUTION:

The original volume of liquid oxygen is 70 liters. To calculate its mass, we must know the density of oxygen as a liquid: 1.141 kilograms per liter. The formula for calculating the mass of a liquid is:

$$mass = density \times volume$$

Next the value for the density of oxygen is plugged in:

1.141 x 70 = 79.87 kilograms (79,870 g)

Once the mass is known, it is easy to calculate the number of moles that this contains:

$$79{,}870 / 32 = 2{,}495.94 \text{ moles.}$$

The next step is to calculate the volume that this mass of oxygen will occupy at 20°C. This is done using what is known as the ideal gas law, $PV = nRT$. In this equation, P is the pressure, V is the volume, $n$ is the number of moles and T is the temperature. R is a value known as the ideal gas constant; this is a figure that features widely in the physical sciences although, for simplicity's sake, we will spare you the detailed explanation. It is sufficient to know that it has a value of 0.082 L.atm/(K.mol). One further thing to note: the value for temperature in this equation must be converted to kelvins (K). 0°C is equivalent to 273 K; the calculation to convert 20°C into kelvins is simply 273 + 20 = 293 K.

In order to find the value for V, the equation must be rearranged to $V = nRT / P$. The values, as we have seen, are:

$n = 2{,}495.94$ mol
$R = 0.082$ L.atm/(K.mol)
$T = 293$ K
$P = 1$ atm

With these, the equation can be solved:

$$2{,}495.94 \times 0.082 \times 293 = 59{,}967 \text{ liters.}$$

Liquid nitrogen has a lower boiling point (–196°C) than liquid oxygen (–183°C) and so vessels containing liquid nitrogen can condense oxygen from air. When most of the $N_2$ has evaporated from a vessel, there is a risk that liquid oxygen remaining can react explosively with organic material. Conversely, liquid nitrogen or liquid air can be oxygen-enriched by letting it stand in air. Atmospheric $O_2$ dissolves in it while $N_2$ evaporates preferentially.

All living organisms, animals, and plants use oxygen present in the air for respiration. This is a process in which food is burned in living cells in the presence of oxygen to release the energy required for all metabolic activities. Medical-grade oxygen gas is used to aid people having difficulty breathing due to inadequate amounts of oxygen such as, for example, hospital patients, underwater divers. and astronauts in space. Oxygen uses in industry mainly involve the melting, welding, and cutting of metals. In oxy-acetylene and oxy-hydrogen blowtorches, oxygen is used to produce the very high temperatures (around 3,000°C, or 5,500°F) needed to melt metals. Liquid oxygen is used as an oxidizer together with liquid hydrogen as fuel in spacecraft to generate the required thrust for a rocket heading into space. NASA's Space Shuttle used an External Tank (ET) containing liquid hydrogen fuel and liquid oxygen oxidizer during lift-off and initial ascent.

# OXYGEN AND REDOX REACTIONS

The oxygen concept introduced by Lavoisier changed chemistry and has proved to be one of the most important principles in the field. Not only is oxygen a vitally important element, but the process by which it forms bonds and ions—the reduction-oxidation or redox reaction—is central to the chemistry of combustion, electrochemistry, respiration, and photosynthesis, and acids and bases.

## Oxygen Rules

We now know that oxygen has six electrons in its outer shell and thus needs two in order to fulfill the octet rule; in other words it has a valency of 2 (see p. 78). This means that when other elements bind with oxygen (oxidation) they do so by donating— either in an ionic or covalent sense— two electrons, and when a compound loses oxygen (reduction) it effectively gets those electrons back.

However, reduction and oxidation have now come to mean more than just losing or combining with oxygen; they are now used to refer to any reactions where electrons are gained or lost. Since you cannot have one without the other, reduction and oxidation are two sides of the same coin—half-reactions that collectively make up the redox reaction. The most important rule to remember is that reduction is the gain, and oxidation the loss, of electrons.

## Redox Redux

Redox reactions include ones we've already met, such as combustion and rusting, neutralization, displacement reactions, and electrochemical reactions. The wide range of processes involving redox reactions means that reduction and oxidation can be defined in three ways.

Reduction can mean gain of electrons, but it can also mean loss of oxygen or gain of hydrogen—all three are equivalent, because they involve net gain of negative charge. So when a zinc cation becomes zinc metal it has been reduced through gain of electrons; when red mercury calx (mercury oxide, $HgO$) is heated until it decomposes into mercury and oxygen, it has been reduced through loss of oxygen; when carbon monoxide ($CO$) and hydrogen gas ($H_2$) are combined to produce methyl alcohol ($CH_3OH$) the carbon monoxide has been reduced through gaining hydrogen.

Oxidation can mean loss of electrons, gain of oxygen, or loss of hydrogen. When sodium and chloride are combined to form the ionically bonded molecule table

## OXIDATION NUMBERS

The oxidation number of an atom or ion is its apparent charge in a compound—the charge it would have if electrons were completely transferred. In an ionic bond the electrons are completely transferred, so the oxidation number of a monatomic ion is the same as its charge (for example, the oxidation number of $Ag^+$ is +1 and of $Ca^{2+}$ is +2). In a covalent bond the electrons are shared but are more attracted to one atom, giving each atom an apparent charge. For instance, in water the electron pairs that make each covalent bond are more attracted to the oxygen atom, giving it an apparent charge of $^{2-}$ and therefore an oxidation number of -2; each hydrogen atom has an oxidation number of +1. Oxidation numbers follow simple rules and allow chemists to predict the ratio by which elements will combine without having to memorize each compound. Some elements, particularly metals, can have different oxidation numbers in different situations; these differing oxidation states are indicated by Roman numerals in brackets: for example, copper(I) has an oxidation number of +1 while copper(II) has an oxidation number of +2.

salt ($Na + Cl \longrightarrow NaCl$), the sodium has been oxidized by losing an electron to the chlorine; when carbon burns it is oxidized to carbon dioxide by gaining oxygen atoms; when the methyl alcohol reaction is reversed ($CH_3OH \longrightarrow CO + 2H_2$), methyl alcohol has been oxidized to carbon monoxide by losing hydrogen.

## Displacement Reactions

In all these reactions, reduction of one substance has been matched by oxidization of the other and vice versa. A good illustration of this is a displacement reaction, such as copper displacing silver from a nitrate solution:

$$Cu(s) + 2AgNO_3(aq) \longrightarrow$$
$$Cu(NO_3)_2(aq) + 2Ag(s)$$

In practice in such a reaction, the silver nitrate is split into ions ($Ag^+ + NO_3^-$) because it is in solution, and the nitrate ions are known as spectator ions because they do not take part in the reaction. What is really happening is that the silver ion is oxidizing the copper (known as an oxidizing agent), while the copper is reducing the silver. The net-ionic equation shows only the active ions:

$$Cu(s) + 2Ag^+(aq) \longrightarrow Cu^{2+}(aq) + 2Ag(s)$$

Breaking this down still further to its component half-reactions makes clear the electron transfer involved in the redox reaction (electrons are $e^-$):

$$Cu(s) \longrightarrow Cu^{2+}(aq) + 2e^- \text{ [oxidation]}$$
$$2Ag+(aq) + 2e^- \longrightarrow 2Ag(s) \text{ [reduction]}$$

# HYDROGEN AND BALLOMANIA

Hydrogen is in many ways the primal element—the first in the periodic table as it was the first created element in nature, produced by the Big Bang. It remains the most common element in the universe, making up the vast majority of the cosmos. Here on Earth it could hold the key to a greener future, although its best-known application has been lighter-than-air flight.

## The Primal Element

According to the Harvard astrophysicist Steven Weinberg, between 70 and 80% of the observable universe is made up of hydrogen; the element accounts for roughly three-quarters of the mass of the universe and makes up more than 90% of all the molecules in existence. Although it was first characterized as a new element by Cavendish, it had previously been produced by medieval alchemists including Paracelsus, which was not surprising as they commonly worked with strong acids and metals—a combination that produces hydrogen. In the 17th century, Frenchmen Theodore Turquet de Mayerne (1573–1655) and Nicolas Lemery (1645–1715) evolved hydrogen by adding iron to sulfuric acid and noted that it was highly flammable, but following Paracelsus they assumed it was a manifestation of sulfur. It was not until Lavoisier repeated and extended Cavendish's experiments with water and achieved its decomposition was hydrogen understood to be an element and given its name.

Cavendish had blown soap bubbles with hydrogen and noted their buoyancy, and Lavoisier was able to measure this property precisely, noting that hydrogen weighed only $\frac{1}{13}$ as much as common air. It did not take long for the wider community to realize an obvious potential application for the new gas. In 1782, Joseph-Michel Montgolfier (1740–1810), who owned a papermaking company with his brother, was prompted to imagine an army of French soldiers invading Gibraltar borne aloft by gas-filled paper bags. The following year the Montgolfier brothers would opt for hot air as their buoyancy aid, but in Paris on August 27th, 1783, academician Dr. Jacques Alexandre Charles (1746–1823) achieved a much faster and higher ascent with a silk bag filled with hydrogen.

## A Road in the Air

In September the Montgolfier brothers caused a sensation at Versailles by sending aloft a hot air balloon carrying a sheep, a duck, and a cockerel. France, with the rest of the world not far behind, was caught in

## COAL OF THE FUTURE

Oxidation of hydrogen, whether through burning or via the safer method of electrochemical combination to generate energy, as happens in a fuel cell, has been widely touted as the energy of the future. Hydrogen can be produced through decomposition of water, and in turn its oxidation produces only water.

As early as 1874 Jules Verne had a character in a novel proclaim: "I believe that water will one day be employed as fuel, that hydrogen and oxygen which constitute it, either singly or together, will furnish an inexhaustible source of heat and light ... Water will be the coal of the future."

the grip of Ballomania. Sir Joseph Banks, president of the Royal Society, was forced to concede that the Montgolfiers' "aerostatic experiment" had "opened a road in the air," marking a new "epoch." The world waited to see who would be the first to send up a man.

In Paris on November 21st the first manned ascent took place in a Montgolfier, as hot air balloons were now known, and just ten days later Dr. Charles and an assistant took to the skies in a hydrogen balloon that incorporated most of the features of a modern balloon, including a wicker basket, impermeable balloon made of silk coated with rubber, a valve for venting gas, and a ballast system. His ascent attracted a crowd of 400,000 people—half of Paris. "With ballooning, science had found a powerful new formula," notes the biographer Richard Holmes, "chemistry plus showmanship equaled crowds plus wonder plus money." Although many applauded the

sheer delight of the new contraptions (Benjamin Franklin, the American ambassador in Paris, recorded that, "Someone asked me—what's the use of a balloon? I replied—what's the use of a newborn baby?") there were doubts as to their utility. Samuel Johnson opined, "I know not that air balloons can possibly be of any use." This was not entirely fair; Joseph Gay-Lussac made a notable ascent in 1804, reaching 23,000 feet (7 km) above Paris in a hydrogen balloon and thereby ascertaining the limit at which air was still breathable. Ballomania tapered off but hydrogen was still used for balloons up until the Zeppelin era, after which helium replaced it as the buoyancy aid of choice.

• On November 21st, 1783, Jean-François Pilâtre de Rozier and François Laurent d'Arlandes made the first recorded manned balloon ascent in a giant "Montgolfier."

# SYSTEMATIC NOMENCLATURE

Among the many legacies of Lavoisier and his French school was the introduction of a new language for chemistry, a scientific language. By clearing away the obscuring thicket of terminology and nomenclature that had built up in haphazard fashion during the evolution of chemistry, the new system brought clarity of definition and clarity of thought; its consequences are still with us today.

## Chemical Babel

Over the millennia, alchemy and industry had built up a rich legacy of colorful and chaotic nomenclature (naming system). Names came from different traditions, languages, and regions. They could derive from locale, method of manufacture, subjective properties (such as taste, smell, consistency, and color), the person who had discovered them, or esoteric factors such as astrology or supposed magical influence. A single chemical could have multiple names, reflecting different historical sources or simply different methods of production. Nitric acid, for instance, was known as spirit of niter when distilled from saltpeter, or as *aqua fortis*. A name like *aqua regia* reflected a whole set of beliefs and assumptions. Terms such as "earth," "oil," and "air" had little specificity or consistency. Names might change completely for the same substance in different phases or in solution.

## Reformation

In the 18th century the increasing pace of discovery of new elements and compounds focused attention on the problem, and attempts were made to reform and standardize the nomenclature. Swedish chemist Torbern Bergman, inspired by the binomial system of botanical nomenclature devised by his countryman Linnaeus, came up with a similar system, and this in turn influenced the French chemist Louis-Bernard Guyton de Morveau (1737–1816). In a paper of 1782 he proposed that chemical names should be short, based on classical roots, and reflect the composition of a substance.

Guyton de Morveau's ideas bore fruit in 1787 with the publication of *Method of Chemical Nomenclature*, by himself, Lavoisier, and two other French chemists, Antoine-François de Fourcroy and Claude-Louis Berthollet.

• Louis-Bernard Guyton de Morveau (1737–1816).

## SOUR GRAPES

Its dependence on Lavoisier's theories meant the new system received a rough ride when it was introduced. However, German and British chemists were forced to learn its principles in order to read Lavoisier's books, and the system caught on. It even survived successive revisions of Lavoisier's theories—for instance, when Davy proved that hydrochloric acid contains no oxygen and thus demolished the Frenchman's theory of oxygen as the acid principle, arguably the name should have been changed. The British accepted the new classical names, but in Germany many of them were translated, so that in German oxygen is still *sauerstoff* ("acid stuff") and hydrogen *wasserstoff* ("water stuff").

The system set out in the *Method* fulfilled Guyton de Morveau's criteria, but was controversial because it was based on Lavoisier's theories—theories regarded as unproven by many chemists,

particularly in Britain and Germany. For instance, the names of mixed bodies (compounds) would be created by combining the names of the simple bodies (elements) of which they were composed, but this depended on Lavoisier's definition of an element and his contention that many well-known substances such as water were actually mixed bodies. Under the new scheme litharge or white lead became lead oxide, while stinking gas became sulfuretted hydrogen gas.

There was no place for phlogiston, and only 33 substances were included as elements (some later proved to be oxides). Recently and newly discovered elements, such as oxygen and hydrogen, would be named for their chemistry rather than any subjective or sociocultural logic, but again, these names reflected Lavoisier's theories. Terminations and word order would show proportions, so, for instance, sulfuric acid was supposed to contain more of the acid principle (oxygen) than sulfurous acid.

For Lavoisier, the reformation of nomenclature was central to the project for a scientific chemistry: *"une langue bien faite est une science bien faite"* ("a well-made science depends on a well-made language"). In the preface to his seminal *Elements of Chemistry* he wrote:

"As ideas are preserved and communicated by means of words, it necessarily follows that we cannot improve the language of any science, without at the same time improving the science itself; neither can we, on the other hand, improve a science without improving the language or nomenclature which belongs to it."

### THE PROBLEM:

French scientist Antoine-Laurent Lavoisier, along with others, is credited with introducing a systematic nomenclature for naming inorganic compounds. The system is still in use today. With the advent of modern organic chemistry, the International Union of Pure and Applied Chemistry (IUPAC)'s system of nomenclature was established and applies to all inorganic, organic, and now biochemical compounds. Mike has recently joined an environmental protection agency and is reviewing his knowledge of hydrocarbon compounds. What are the simplest hydrocarbons in petrol and how are they named?

### THE METHOD:

Hydrocarbons are organic compounds, meaning they have carbon-hydrogen (C-H) bonds and also carbon-carbon (C-C) bonds, where carbon is bonded to other carbon atoms. Both bonds are fundamental building blocks of the physical world.

Carbon can bond to four other atoms, similar or different. It can also form double and triple bonds with itself and one or two other elements. In terms of the elements bonded, the simplest group of organic compounds is the hydrocarbons (HCs) which contain just carbon and hydrogen atoms. A saturated hydrocarbon, or alkane, is a hydrocarbon in which all of the carbon-carbon bonds are single bonds (–C–C–C–C)– a form known as a carbon chain. The alkanes differ from one another only by the number of carbon atoms in the carbon chain. The simplest member of the alkanes is methane ($CH_4$), which has one carbon atom and four hydrogen atoms. The next is ethane ($C_2H_6$), which has two carbon atoms. The total number of hydrogen atoms can be calculated doubling the carbon number and adding 2. Ethane therefore has six hydrogen atoms.

## THE SOLUTION:

Starting from a carbon number of 1, and remembering that carbon can bond to 4 atoms, the names and structures of the alkanes are respectively:

methane ($CH_4$),
ethane $CH_3CH_3$ ($C_2H_6$),
propane $CH_3CH_2CH_3$ ($C_3H_8$),
butane $CH_3CH_2CH_2CH_3$ ($C_4H_{10}$),
pentane $CH_3CH_2CH_2CH_2CH_3$ ($C_5H_{12}$),
hexane $CH_3(CH_2)_4CH_3$ ($C_6H_{14}$),
heptane $CH_3(CH_2)_5CH_3$ ($C_7H_{16}$),
octane $CH_3(CH_2)_6CH_3$ ($C_8H_{18}$),
nonane $CH_3(CH_2)_7CH_3$ ($C_9H_{20}$),
decane $CH_3(CH_2)_8CH_3$ ($C_{10}H_{22}$).

When hydrocarbons burn, they release water, carbon monoxide (partial combustion) and carbon dioxide (complete combustion), and energy for powering vehicles and electricity generation. Hydrocarbons are also released into the atmosphere as polluting gases from incomplete combustion of fossil fuels as well as by evaporation.

When hydrocarbons combine with nitrogen oxides ($NO_x$) in a vehicle exhaust gas they can, under ultraviolet light from sunlight, produce toxic ozone (trioxygen). This collects at ground level, contributing to pollution and forming a key component of photochemical smog, a major health problem in many cities around the world. The presence of low-level ozone in cities is unrelated to the stratospheric ozone layer in the upper atmosphere. Catalytic converters fitted to a vehicle exhaust can help reduce noxious emissions.

• Diagrammatic model of an octane molecule. The name of this hydrocarbon immediately conveys an important structural property: it is an alkane with eight carbon atoms.

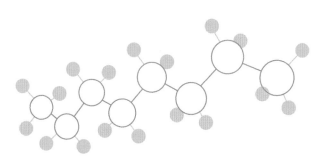

# Atoms and Ions

Lavoisier's chemical revolution changed the
way chemistry was done and inspired a new
generation of scientists, yet the young science still
lacked many of the attributes Newton had brought
to physics, such as simple mathematical principles
and laws that would make it a truly quantitative
science. This chapter explains these principles and
laws, and tells how they came to be discovered,
introducing along the way an electrifying new tool
for the analysis of matter.

# ATOMIC WEIGHTS AND ATOMIC THEORY

While chemistry was now firmly established as a science, many aspects remained mysterious and chaotic. Which substances were elements? What were elements made of? How did they come together to make compounds and how could chemists work out the formulae of those compounds? The microcosmic world of matter seemed impenetrable, but an insight of breathtaking simplicity would bring it into focus.

## Definite Proportions

Atomism had been revived in the 17th century by corpuscularians like Boyle, and most of the scientific world had come to accept Isaac Newton's speculations on the issue in the *Principia*. Newton conceived of matter as discrete, indivisible particles that interacted not because of their shape (as the ancient Greeks had argued), but via forces of attraction and repulsion similar to gravity but operating on a much smaller scale. In other words, the microcosmic world of atoms was a mirror of the macrocosmic world of planets and moons.

This made for a plausible theory but in the absence of the powerful technology needed to explore the atomic world directly, was of little practical use until the advent of chemical atomism. In 1788 French chemist Joseph-Louis Proust (1754–1826) discovered the law of constant composition, also known as the law of definite proportions. Previously it had been accepted that the composition of a compound could vary, so that, for instance, some bodies of water might have more oxygen than others, or that making copper carbonate in one way would result in a compound with a higher proportion of copper than, say, obtaining it from a mineral deposit. Proust's careful analysis showed that this was not the case; the composition of copper carbonate was identical however it was made and wherever it was obtained. The same rule applied to other compounds—they all consisted of elements in definite simple ratios by weight, and what was more, the proportions were basically integers.

## A New System

John Dalton (see pp. 122–123) realized that Proust's law could be explained by atomic theory: compounds must be formed through the combination of discrete particles, and these particles must vary in weight by integer multiples. In 1808 Dalton published *A New System of Chemical Philosophy*, the founding text

# THE SIMPLICITY OF NATURE

Although the terms are sometimes used interchangeably, a distinction can be made between atomic mass and atomic weight (although both are measured in atomic mass units or amu). Atomic mass is the same as the mass number of an atom—the sum of protons and neutrons in its nucleus. As has already been mentioned, mass of a proton or neutron is defined as $\frac{1}{12}$ the mass of an atom of carbon-12 (an isotope of carbon with six protons and six neutrons). Elements in nature generally exist as a mixture of different isotopes with slightly different atomic masses; atomic weight is the weighted average of atomic masses of the various isotopes. For instance, carbon mostly exists in nature as carbon-12, but a small proportion of any body of carbon will also contain atoms of carbon-13 and carbon-14, which have atomic masses of 13 and 14 respectively. The average atomic mass of carbon is therefore 12.011, and this is the atomic weight.

of chemical atomism. In it he explained that each element has its own characteristic atoms, distinguished by their relative weights. He did not speculate as to the other properties of atoms—there was no way to examine these—but careful quantitative chemistry meant it was possible to determine the relative weights of atoms of different elements.

Hydrogen was the lightest element known, so Dalton assigned it an atomic weight of 1, and calculated other elements accordingly. Analysis of water by Lavoisier and others had shown that its components were oxygen and hydrogen in a ratio by weight of 8:1. Dalton assumed that nature kept things as simple as possible, therefore the most likely formula for water was the simplest—one atom of hydrogen to one atom of oxygen. Accordingly oxygen must have a relative or atomic weight of 8. Using these weights he could work out the weights of other elements. For instance, he knew that the compound he called carbonic oxide (carbon monoxide) was composed of carbon and oxygen in the ratio by weight of 3:4. Given that oxygen had an atomic weight of 8, carbon must therefore have an atomic weight of 6. In practice Dalton's assumptions were often wrong and this threw off his calculations, so that few of his atomic weights proved accurate, but he had established chemical atomism and set chemists on the road to quantifying their science.

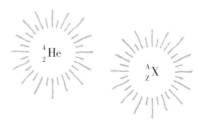

· Notation showing atomic number and mass number; on the right is the generic version for element X, on the left is helium, with a mass number (A) of 4 and an atomic number (Z) of 2.

# John Dalton

**A humble man from Britain's provinces with little formal education who was isolated from the centers of intellectual power, John Dalton became a famous figure despite himself. His discoveries helped to advance the development of chemistry, while his career marked an important stage in the evolution of science in general.**

## Dalton's Law

Born into a family of devout Quakers in Cumbria, a rural district in the north of England, John Dalton (1766–1844) was destined to be an outsider in the world of 19th-century British science, a world of privilege, gentleman-amateur scientists and an establishment firmly centered on the old guard of the Royal Society in London.

As a Quaker, a dissenter, Dalton was barred from attending the great universities even if he could have afforded it. Instead his education was limited to the village Quaker school, which he attended until aged 12, whereupon he started teaching there. Later he became a headmaster, a lecturer and finally a private tutor, moving to Manchester where he was taken up by

supportive fellow Quakers who encouraged his scientific research.

Dalton's initial interests lay in the field of meteorology; he kept meticulous meteorological records every day of his life, for 57 years, up until the day he died. This led him to an interest in water in all its phases; he determined, for instance, that the density of water varies with temperature (water is most dense at 39°F [4°C]). Studying water vapor led on to pneumatic chemistry in general, and Dalton soon adopted Boyle's corpuscularian view of gases, becoming a hardline atomist. Even after accepting his atomic theory (see pp. 64–65), most of the chemical world reserved judgement on whether atoms were genuine physical objects or simply useful theoretical entities. Dalton had no such qualms. "All bodies of sensible magnitude, whether liquid or solid," he wrote, "are constituted of a vast number of extremely small particles, or atoms, of matter bound together by a force of attraction ..." He was an equally fervent believer in the law of conservation of matter: "No new creation or destruction of matter is within the reach of chemical agency. We might as well attempt to introduce a new planet into the solar system, or

• **ONE IN THE EYE**

*Dalton was intrigued by his own color-blindness, and lectured on the subject. For a period the condition came to be known in Britain as Daltonism. He specified that on his death his eyes were to be preserved and dissected.*

## SCIENCE IN PROVINCIAL BRITAIN

Dalton was one of a new breed; men of no station or family wealth who made their way in life as professional scientists, at a time of growing tension between the provinces and the gentleman-amateur scientific establishment based in London and centered on the Royal Society. He played an important role in setting up the British Association for the Advancement of Science (or BA, as it was known), which was conceived as an alternative to the Royal Society and reflected the growing professionalism of the field. The BA was established in 1831 by founder members including the Scottish physician (and biographer of Newton) Sir David Brewster, who complained that Britain's "scientific institutions have been discouraged and even abolished." Its first meeting was at York and thereafter it met annually, mostly outside London, and was the forum at which major advances in British science were announced.

to annihilate one already in existence, as to create or destroy a particle of hydrogen."

This atomist conception of gases led him to formulate what is now known as Dalton's law, which he published in 1801. Also known as the law of partial pressures, it says that in a gaseous mixture each component gas exerts pressure independently of the other gases, so that the total pressure is the sum of the pressures of each component. This only holds true for ideal gases and thus a corollary of the law was that in gaseous mixtures, such as the atmosphere, the components have no chemical interaction with one another. Since it was widely believed at the time that the atmosphere was a compound of its various airs, this caused a stir.

## A Very Singular Man

His theory of atomic weights brought greater fame, but as an outsider from the provinces who was reluctant to join the Royal Society, which he considered a body of amateurs, Dalton was not always well thought of by contemporaries. Humphry Davy (see pp. 142–143) described Dalton as "a very singular Man ... He has none of the manners or ways of the world." His brother John Davy was uncharitable: "Mr. Dalton's aspect and manner were repulsive. There was no gracefulness belonging to him. His voice was harsh and brawling; his gait stiff and awkward; his style of writing and conversation dry and almost crabbed. In person he was tall, bony, and slender." Humphry cast doubt on Dalton's ability as an experimental scientist, commenting archly that, "He was a very coarse Experimenter & almost always found the results he required." Despite this, Dalton gained international renown and 40,000 people attended his funeral.

# Accurate Atomic Weights

## THE PROBLEM:

Dalton formulated his atomic theory of matter and estimated the atomic masses of elements based on hydrogen having an atomic mass of 1. As we've seen, the atomic weight of elements is now based on the mass of the carbon-12 ($^{12}C$) isotope. Jon is working out his company's carbon footprint. To achieve this he first needs to calculate the accurate atomic masses of carbon and hydrogen. How does he go about doing this?

## THE METHOD:

According to the official definition, the relative atomic mass of an element from a specified source is the ratio of the average mass per atom to $^1/_{12}$ of the mass of an atom of $^{12}C$. To calculate the atomic masses of carbon and hydrogen, Jon must first find the masses of the natural isotopes of carbon and hydrogen and also their respective relative abundance.

## THE SOLUTION:

From National Institute of Science and Technology (NIST) tables, the two naturally occurring isotopes of carbon, $^{12}C$ and $^{13}C$, have masses of 12.000000 and 13.003355 with respective relative abundance of 98.90% and 1.10%.

The two naturally occurring isotopes of hydrogen, $^1H$ and $^2H$, have masses of 1.007825 and 2.0140 with respective relative abundance of 99.985% and 0.015%. As we saw in Exercise #7 (pp. 62–63), the formula for atomic weight is:

(mass x abundance) + (mass x abundance)

After plugging in the values we have for the two isotopes of carbon, the calculation is straightforward:

$$(12 \times 0.9890) + (13.003355 \times 0.011)$$

$$11.868 + 0.1430369$$

$$= 12.0110369$$

Applying the same process, the atomic mass of hydrogen is:

$$(1.007825 \times 0.99985) + (2.0140 \times 0.00015)$$

$$1.0076738 + 0.0003021$$

$$= 1.0079759$$

Using NIST tables, the calculated atomic weights of carbon and hydrogen that Jon obtained are thus respectively 12.01104 and 1.00798. The official 2007 International Union of Pure and Applied Chemistry (IUPAC) values are 12.0107(8) and 1.00794(7), where the number in brackets indicates the uncertainty in the last digit of the atomic weight. Discrepancies between the values at this degree of accuracy are inevitable. In the field of biochemistry and molecular biology, the term "dalton" (Da) is often used for the mass of molecules where 1 Da is the mass of a hydrogen atom. Because proteins are large molecules, their masses are often quoted in kilodaltons (kDa). The largest known proteins are the titins, a component of the muscle sarcomere, with a molecular mass of almost 3,000 kDa.

• Illustration showing the structures of the nuclei of two isotopes each of carbon and hydrogen. The neutrons are represented here in white; the protons are in green.

Hydrogen nucleus
(1 proton)

Carbon-13 nucleus
(6 protons + 7 neutrons)

Carbon-12 nucleus
(6 protons + 6 neutrons)

Deuterium nucleus
(1 proton + 1 neutron)

# MOLES AND AVOGADRO'S NUMBER

Atoms and molecules are extremely small and weighing or counting them directly is not feasible. This is where the mole comes in, giving chemists a way of relating atomic weights to real-world weights and measures and determining the actual formulae of compounds. The mole is the bridge between the microscopic and macroscopic worlds.

## Counting by Weighing

A mole is an amount of a substance that contains as many particles as there are atoms in exactly 12 grams of carbon-12. The particles can be anything—atoms, molecules, ions, electrons—but they must be specified. The experimentally determined number of atoms in 12 grams of carbon-12 is known as the Avogadro number, named after the 19th-century scientist who effectively originated the concept of the mole; it is $6.0221367 \times 10^{23}$, which is 602 billion trillion, or 602 with 21 zeros after it. The mole expresses the atomic weight of an element in grams; namely, in quantities chemists can work with in the real world. Moles allow us to count by weighing.

The mole is not simply atomic weight expressed in grams; it can also be the molecular, or formula, weight expressed in grams. Formula weight is the sum of the atomic weights of atoms in a compound (molecular weight is the same thing, but for covalently bonded molecules only; formula weight is a more general term that covers ionic compounds too). For instance, the atomic masses of the atoms that make up a molecule of water are 1 amu each for the two hydrogen atoms and 16 amu for the oxygen, giving a sum total of 18 amu, so the formula mass of water is 18 amu. A mole of water therefore weighs 18 grams. (Actually the atomic weights of these elements are not quite round numbers because of isotopes, so the formula weight of water is $(2 \times 1.0079) + 15.999 = 18.015$ amu.)

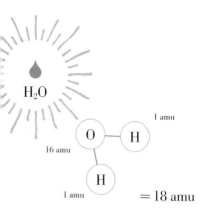

$H_2O$

1 amu

O — H

16 amu

H

1 amu

$= 18$ amu

## Moles in Action

The mole is a powerful tool for chemists. Say you have 22.99 g of sodium (Na) and you want to make table salt by combining it with chlorine (Cl), but you don't want to waste reactants by having too much of one or the other. How do you know how much chlorine to use? Handily the atomic weight of sodium is 22.99 amu, so you know you've got 1 mole of sodium. Given that you also know the formula of table salt is NaCl, you know that you need one atom of chlorine for each atom of sodium, or 1 mole of chlorine for each mole of sodium. The atomic weight of chlorine is 35.453, therefore you need 35.453 g of chlorine. In practice, no reaction is 100% efficient so not every particle of reactant will react (plus chlorine is diatomic in its gaseous form), but you get the idea. Weighing gases is laborious, so helpfully it is possible to convert the mole concept into measures of liquid and gaseous volume (see box).

The calculations also work in reverse. Supposing an analytical chemist knows that he produced 400 grams of compound Z from 300 grams of reactant X

### MOLE EXTRAS

Terms related to the mole concept are molar mass and molar volume. The molar mass of a substance is the mass of a mole of that substance, expressed as grams/mole ($g\ mol^{-1}$). The molar volume ($V_M$) of a substance is the volume occupied by a mole of that substance, which depends on its density. This in turn depends on temperature and pressure, although the density of liquids varies little, so $V_M$s for liquids at room temperature and sea level are useful across a wide range. For gases, $V_M$ depends on the precise temperature and pressure; at standard temperature and pressure the $V_M$ for any gas is 22.415 $dm^3$.

and 100 grams of reactant Y. The percentage composition by weight of Z would be 75X:25Y, suggesting a formula of $X_3Y$. But the chemist knows that the atomic weights of X and Y are 50 and 25 amu respectively, so he can calculate that he has combined 6 moles of X with 4 moles of Y, suggesting an empirical formula (the lowest integer ratio of elements in a compound based on the percentage composition) of $X_3Y_2$. The formula weight of the compound Z according to this would be 200 amu, so the chemist would deduce that he has produced 2 moles of Z. If separate research showed that Z had a formula weight of 400, the molecular formula would be $X_6Y_4$.

# Avogadro the Unheralded

**The man who made the vital link between the micro- and macroscopic worlds with a daring conceptual leap was ignored and dismissed in his own time. Despite training as a lawyer Amedeo Avogadro took private lessons in maths, chemistry, and physics before embarking on a career in science.**

## Holding out for a Hero

Dalton's work with atomic weights had run into one, seemingly insurmountable, obstacle. Although it was possible to determine relative proportions of elements in a compound, there was no way to relate this definitively to the formula of that compound. For example, Dalton had assumed on grounds of simplicity that water was a 1:1 mix of hydrogen and oxygen, and this faulty premise had led to a faulty conclusion: he had got the atomic weight of oxygen wrong. This in turn undermined his whole system. It took the insights of Amedeo Avogadro (1776–1856) to show the way forward, leading to the concept of the mole, which made it possible to determine actual atomic weights and thereby empirical formulae (see pp. 136–137).

The man who made this possible was an unassuming minor nobleman from northern Italy who practised law until 1800, when he began his scientific education. Although he became primarily a mathematical physicist, Avogadro recognized no boundaries between the sciences. The rest of the chemical community was more blinkered, perhaps, as his ideas made little impact (see box), partly because of intellectual snobbery that a mere mathematician should advance radical chemical theories.

## Avogadro's Law

Avogadro's breakthroughs were based on two discoveries by French chemist Joseph Gay-Lussac (1778–1850): firstly, that all gases expand equally as temperature increases, and secondly the law of combining volumes. This said that the volumes of gases that react with one another, or are produced in a chemical reaction, are in the ratios of small integers (for example, 2 volumes of hydrogen + 1 volume of oxygen $\longrightarrow$ 2 volumes steam; 1 volume hydrogen + 1 volume chloride $\longrightarrow$ 2 volumes hydrogen chloride). Although Dalton himself was unable to appreciate it, Gay-Lussac's law was the direct counterpart of Proust's law of constant composition and a powerful confirmation of his atomic theory.

Avogadro, however, did make this leap, boldly asserting that Gay-Lussac's first discovery meant that "equal volumes of all gases, under the same conditions of temperature and pressure, contain the same number of smallest particles," now known as Avogadro's law. Note Avogadro's careful use of the term "smallest particles"; he was the first to use the word "molecules," and his "molecular hypothesis" suggested

• Unflattering contemporary portrait, made in the year of his death, of Lorenzo Romano Amedeo Carlo Bernadette Avogadro di Quaregna e Cerreto.

## PROPHET WITHOUT HONOR

So why the long silence? The answer probably lies in a combination of geographical and intellectual isolation, scientific snobbery, and the dominant paradigms of the day. Avogadro was considered a provincial if he was considered at all. His base of Turin was far from the centers of scientific power at the time, and the upheaval of the Napoleonic era in this region (which saw him lose his academic post for several years) probably did not help. A reputation as a poor experimentalist meant that other chemists did not take him seriously, and he damaged his own cause by failing to support hypotheses with enough hard data, especially when he tried to extend his speculations about diatomic elements to the world of solids (where he was proved wrong). Avogadro's contention that oxygen and hydrogen were diatomic also told against him, because at the time the dominant theory was the electrochemical dualism of Jöns Berzelius (see pp. 134–135), which said that atoms of the same element would repel each other like ions bearing the same charge.

that gases such as oxygen and hydrogen might be diatomic molecules. Using his law and the law of combining volumes, it was possible to work out that hydrogen and oxygen must be combining to form water in the ratio 2:1, and therefore that the molecular formula of water must be $H_2O$. This in turn made it possible to calculate the correct atomic weights of the elements from the percentage composition by weight. Avogadro had succeeded in reconciling the discoveries of Gay-Lussac and Dalton and opened the way for the proper quantification of chemistry.

The real mystery is why his brilliant idea made no impact at the time. Not until the Karlsruhe Conference of 1860, when Avogadro was already dead, did his compatriot Stanislao Cannizzaro (1826–1910) demonstrate the power of Avogadro's hypothesis and force the

scientific community into a long-overdue reappraisal. Instead, needless debate and confusion over the true atomic weights and molecular formulae of oxygen, hydrogen, water, and many other substances continued.

• Avogadro's law: mass and molecular formula of a gas may vary, but the volume occupied by 1 mole at standard pressure and temperature is a constant at 22.4 liters/mol.

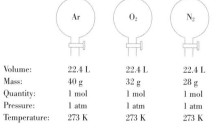

| | Ar | $O_2$ | $N_2$ |
| --- | --- | --- | --- |
| Volume: | 22.4 L | 22.4 L | 22.4 L |
| Mass: | 40 g | 32 g | 28 g |
| Quantity: | 1 mol | 1 mol | 1 mol |
| Pressure: | 1 atm | 1 atm | 1 atm |
| Temperature: | 273 K | 273 K | 273 K |

# 17 Avogadro and the Mole

### THE PROBLEM:

Returning to the gas laws, Avogadro's law states that the volume occupied by an ideal gas in a container is proportional to the number of moles (or molecules) present, and this law can even be applied to water vapor. At the other end of the temperature scale, Wayne is pondering the well-known uniqueness of snowflakes under the microscope. He wonders how many water molecules there are in a typical snowflake.

### THE METHOD:

The mole (symbolized by "mol") is defined as the amount of substance that contains as many particles (atoms, molecules, ions, or electrons) as there are atoms in 12 grams of the isotope carbon-12 ($^{12}C$). It follows then that 1 mole of pure $^{12}C$ atoms has a mass of exactly 12 grams. The number of atoms or molecules contained in 1 mole of a pure substance is known as the Avogadro constant, which is 6.022142 x $10^{23}$ mol$^{-1}$. Thus, a mole of any pure substance has its mass in grams equal to that of its atomic, or molecular (if the substance is a compound), mass. For the calculation, Wayne needs the number of molecules in 1 mole of water and also the atomic weights for hydrogen (H = 1.01) and oxygen (O = 16).

## THE SOLUTION:

Snowflakes are made of water ($H_2O$) and weigh 1 milligram. In order to calculate how many moles are present in a single snowflake, the molecular mass of $H_2O$ must be calculated, from which the mass of 1 mole of water is derived:

$$(2 \times 1.01) + 16.00 = 18.02$$

Therefore, 1 mole of $H_2O$ weighs 18.02 grams.

The next step is to determine the number of $H_2O$ molecules in 1 gram of water. Using Avogadro's number, we know that 1 mole of $H_2O$ contains $6.022 \times 10^{23}$ molecules. Therefore, in 1 gram of $H_2O$, the number of molecules is:

$$(6.022 \times 10^{23}) / 18.02 = 3.34 \times 10^{22}$$

A single snowflake weighs just 1 milligram (a thousandth of a gram) so will therefore contain $3.34 \times 10^{19}$ molecules of $H_2O$.

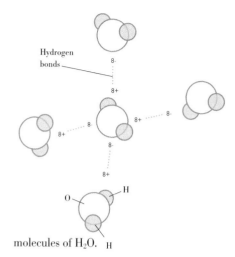

molecules of $H_2O$.

• A model demonstrating the hydrogen bonds (H-bonds) between water molecules ($H_2O$). A molecule of $H_2O$ can bond with up to four other water molecules, as demonstrated in the diagram. H-bonds are influential in determining the properties of water: their relative strength, and the high number of such bonds that can be formed by a water molecule (relative to its mass), contribute to water's high boiling point. They also contribute to the hexagonal lattices that form the distinctive crystalline patterns of snowflakes.

# IONS AND CHARGES

The advent of electrochemistry opened up an exciting new field of science. Understanding this brave new world requires a quick review of the basics of ions and charges, including the apparently complicated but in fact quite logical system of names for ions of different oxidation states.

## Ionic Basics

As we saw in Chapter 2 (see "Covalent and ionic bonds," pp. 78–79), atoms seek to gain energetic stability by achieving a stable electronic configuration, losing or gaining electrons from their outer valence shells. When an atom completely donates or receives one or more electrons (rather than just sharing them) there will be a mismatch between the number of protons it has and the number of electrons, and this will result in the atom acquiring a positive or negative charge, at which point it becomes known as an ion. Atoms that form ions lose or gain them following the octet rule, according to which they seek to become isoelectronic (to have the same configuration of electrons) with the noble gas that is nearest in the periodic table. For instance, in the formation of table salt (NaCl), sodium donates an electron to chlorine, so that the sodium atom becomes a cation with a charge of +1 that is isoelectronic with neon, while the chlorine atom becomes an anion with a charge of -1 that is isoelectronic with argon. Charges are indicated with a superscript after the elemental name, in contrast to numbers of atoms in a molecule which are indicated by a subscript (see the diatomic ion of mercury, below, for an example).

Ionic compounds form through electrostatic attraction between positively and negatively charged ions; this attraction constitutes a ionic bond. A characteristic form of ionic compound is a salt, formed when an acid reacts with a base, usually a metal. Metal salts typically form crystal lattice structures.

## Species Diversity

"Species" is a general term used to describe types of ion, covering the various categories, which include monatomic and polyatomic ions. The species of ion an element or compound can give rise to is governed by periodic law (see pp. 154–155). Amongst monatomic ions, for instance, alkali metals form cations with a charge of +1, alkaline earth metals +2 cations, halogens -1 anions, oxygen and sulfur -2 anions and nitrogen and phosphorus -3 anions. Anions take the suffix "-ide," so that the anions of chlorine, fluorine, sulfur, oxygen, nitrogen and phosphorus are named chloride, fluoride, sulfide,

## THE CRISSCROSS RULE

The sum of charges in an ionic compound must be 0, and this allows you to work out the charges of the component species if you know the molecular formula, by using the crisscross rule: the subscript of each ion tells you the superscript of the other. For instance, aluminum oxide has the formula $Al_2O_3$, so the charges of the two ion species are $Al^{3+}$ and $O^{2-}$ respectively. It's important to remember that a sub- or superscript of 1 is never shown but understood to be present, and that superscripts are reduced by the lowest common denominator.

oxide, nitride, and phosphide ions, respectively.

The transition metals can have different oxidation states, and this means they can form ions with different positive charges, where the charge is equal to the oxidation state. The convention is to indicate the oxidation state and therefore the charge by Roman numerals in brackets, but naming conventions also indicate oxidation state, with the suffix "-ous" indicating the ion with the lower oxidation state. Thus iron(II), $Fe^{2+}$, is ferrous, while iron(III), $Fe^{3+}$, is ferric. Similarly copper(I), $Cu^+$, is cuprous, while copper(II), $Cu^{2+}$, is cupric. Mercury is unusual in forming diatomic cations in one of its oxidation states—mercury(I) or mercurous, $Hg_2^{2+}$; each of the cations has a single + charge, giving the diatomic ion an overall charge of +2.

Apart from mercury there are many other polyatomic species, most of them oxyanions (oxygen containing anions).

Oxyanions take the suffix "-ate," except where a variant with fewer oxygen atoms exists, in which case it ends with "-ite." Hence $SO_4^{2-}$ is sulfate, but $SO_3^{2-}$ is sulphite. Other important polyatomic anions include hydrogen carbonate, aka bicarbonate ($HCO_3^-$), nitrate ($NO_3^-$) and nitrite ($NO_2^-$), hydroxide ($OH^-$), cyanide ($CN^-$), and peroxide ($O_2^-$).

• Diagrammatic view of a mass of solid salt (left), dissolving in water; each $Na^+$ and $Cl^-$ ion attracts water molecules to create a hydration shell.

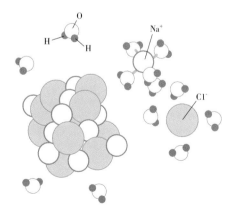

# Jöns Jacob Berzelius

**After Lavoisier the next great figure in chemistry was the Swede, Jöns Jacob Berzelius, who made many discoveries and advanced the theory and practice of his science. After exploring the new art of electrolysis, he went on to perfect techniques of quantitative chemistry, discover new elements and compounds, devise a system of notation and effectively control chemistry in Europe.**

## Enraptured with the Pile

Born in Väversunda in Sweden, Jöns Jacob Berzelius (1779–1848) overcame early difficulties in accessing education by reading every chemistry textbook he could lay his hands on. Although he trained as a doctor so that he could support himself, his true passion was chemistry—particularly the new field of electrochemistry that had been sparked into life by Volta's invention (see box). Historian of science William Burns describes Berzelius as "one of the innumerable European scientists enraptured with the possibilities of the voltaic pile in the early nineteenth century."

In 1803 Berzelius stuck electrodes into a solution of a neutral salt and noted that what he called its "acid" component accumulated around the positive pole and its "base" around the negative pole. A few years later Humphry Davy succeeded in using electrolysis to isolate sodium, potassium, and the alkaline earths (see p. 142),

confirming for Berzelius that the voltaic pile had uncovered the fundamental principle of chemistry. Accordingly he formulated a "dualistic theory"—"In arranging the bodies in order of their electrical nature, there is formed an electro-chemical system which, in my opinion, is more fit than any other to give an idea of chemistry"—classifying all substances according to whether they are electropositive or electronegative. Salts, Berzelius claimed, were formed when electronegative substances acted as acids and electropositive ones acted as bases. He subscribed to Lavoisier's belief that oxygen is the acid principle; indeed he insisted that the acids and bases that make up salts must both be oxides, refusing to accept that chlorine and iodine were elements until the 1820s.

## With Scrupulous Precision

Berzelius eagerly accepted Dalton's work on the principle of combinations in simple proportion, calling it one of "greatest steps that chemistry has made toward its perfection as a science." He dramatically improved the determination of atomic weights and molecular

formulae by attaining new standards of accuracy in the quantification of chemistry, a goal he pursued "with the most scrupulous precision." Berzelius would go on to prepare, purify, and analyze more than 2000 substances, including several new elements. He discovered selenium (1817), thorium (1828), and silicon (1824), while his team isolated

## THE VOLTAIC PILE

The invention that revolutionized chemistry and many other sciences was relatively simple—a stack of alternating silver and zinc discs interleaved with card soaked in brine. This primitive battery, known as the voltaic pile after its creator Alessandro Volta (1745–1827), was able to generate enough voltage for electrolysis. On learning of the new device, William Nicholson and Anthony Carlisle immediately constructed their own, using it to decompose water and printing the results in Nicholson's eponymous journal before Volta's 1800 paper describing the device had even been published.

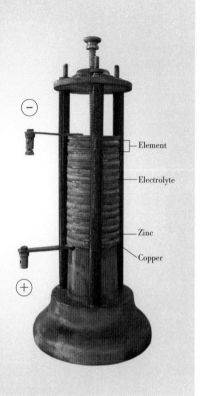

• Replica of an early voltaic pile, with alternating copper and zinc discs sandwiching brine-soaked card. Each "sandwich" is one "element." The brine acts as an electrolyte.

lithium (1818) and vanadium (1830). However, his dualistic theory led him to reject Avogadro's theories about diatomic molecules, and this resulted in much confusion over the atomic weights and formulae of some important elements (especially gases).

Among other achievements, he established that organic compounds follow the same rules of proportional composition as inorganic ones, and helped characterize some important phenomena in organic chemistry, coining the terms "catalyst," "protein," and "isomerism." Perhaps his most remarkable achievement,

however, was the way in which he seized control of his science. "Berzelius's domination of European chemistry from about 1820 is an astonishing phenomenon," notes Burns, "given the marginality of his base at Stockholm to the centers of European science." Through writing a constantly updated standard textbook, and editing a yearbook that was required reading, Berzelius became the gatekeeper for the field. In later years, as he became increasingly hidebound, obstructive, and bitter at being sidelined, this was to prove problematic.

# CHEMICAL NOTATION

The dream of 19th-century chemists was to bring to their science
the same rigor and precision as mathematics, in similar fashion to the
way that Newton had transformed physics. Any modern student
of chemistry knows that they succeeded, for nothing is now more
emblematic of the topic than the chemical equation. Bridging this gap
was one of Jöns Berzelius's lasting achievements.

## The Necessity of Signs

"When we endeavor to express chemical proportions," Berzelius wrote in 1814, "we find the necessity of chemical signs." The year before he had begun to put together a new system of notation to meet this need. As with nomenclature (see pp. 114–115), previous methods of notation had been haphazard and inconsistent, reflecting the uneven development of chemistry over thousands of years and across many cultures and languages. Alchemists had used symbols dense with meaning, often derived from occult or esoteric sources such as astrology, but new substances and the new understanding of elements—as set out in Lavoisier's system of nomenclature—called for a new approach.

John Dalton had constructed his own, rather elaborate, system with a series of simple diagrams. Although utilizing some symbols that are now familiar (his symbol of hydrogen, for instance, now seems eerily prescient), his approach had obvious flaws.

Berzelius, in his 1814 "Essay on the Cause of Chemical Proportions, and on some circumstances relating to them: together with a short and easy method of expressing them," explained them concisely: "Chemical signs ought to be letters, for the greater facility of writing, and not to disfigure a printed book . . . I shall take therefore for the chemical sign, the initial letter of the Latin name of each elementary substance: but as several have the same initial letter, I shall distinguish them in the following manner . . ." He then went on to explain that he would use the first two letters, and if these were common to two elements, the initial and "the first consonant which they have not in common." Thus sulfur = S; silicon = Si; stibium (the Latin name for antimony) = St; and stannum (tin) = Sn. The periodic table on page 149 shows the one or two letter combinations for every element. This system appealed to printers because they could use letters they already had, instead of having to make new ones. It soon became the universal standard.

In the modern version of chemical notation, subscripts after the element symbol represent the number of atoms of that element in the molecule, while superscripts are used to represent positive or negative charge, in the case of ions. Sub- and superscripts before the element symbol represent atomic number and mass, respectively. The atomic mass superscript thus also indicates the isotope of the element, so $^{12}C$, for instance, is carbon-12.

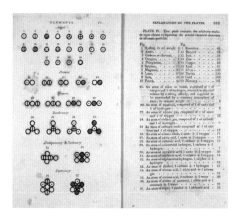

• Dalton devised a new system of notation for the elements but it was unwieldy and would have required printers to develop new typefaces for the symbols; it never caught on.

## Formula for Success

In the new system, compounds could be easily represented by putting symbols together, and Berzelius introduced the convention of having each sign represent one volume, or mass, of a substance, with multiples represented by coefficients (numbers in front of the symbols). Now chemical reactions could be written like mathematical ones. In an equation the two sides are separated by an arrow $\longrightarrow$, which shows the direction from reactants to products. Many chemical reactions are reversible: they can go both ways, even though one direction may proceed faster than the other. Eventually the two will reach a state of equilibrium. In such cases a two-way arrow is used: $\rightleftharpoons$

The law of conservation of matter has an important consequence for chemical equations: because atoms cannot be created or destroyed, there must be the same number of atoms on one side of the equation as on the other. In other words, chemical equations must balance. An equation is incorrect if it does not

meet this criterion. For instance, consider the equation for hydrogen + oxygen = water. Since both hydrogen and oxygen are diatomic, you might write:

$$H_2 + O_2 \longrightarrow H_2O$$

But this equation does not balance because there are two oxygen atoms on the left and only one on the right; just as in algebra, we must multiply the appropriate coefficients by the lowest common factor to balance the equation:

$$2H_2 + O_2 \longrightarrow 2H_2O$$

This is known as balancing by inspection.

# ELECTROLYSIS

Electrolysis means "breaking apart with electricity"; with the advent of the voltaic pile, this phenomenon became a powerful new tool in the hands of analytical chemists. An electrolytic or electrochemical cell can be used to separate ions, produce redox and displacement reactions, break apart compounds, and isolate pure elements.

## The Electrolytic Cell

Electrolysis is a means of producing chemical changes by the passage of an electrical current, through reactions at electrodes immersed in an electrolyte. Electrodes are solids, usually strips of metal, which are connected to a source of electricity such as a battery or voltaic cell. Just as the battery has positive and negative terminals, so the electrodes are positive or negative. The positive electrode is called the anode, and the negative electrode is the cathode. An electrolyte is a conductive solution or liquid with ions that can move and carry charge. A typical electrolyte is brine (salt solution). In brine, sodium chloride dissociates to give sodium cations and chloride anions, and these are able to migrate through the solution when attracted toward the electrodes. Molten salt is also an electrolyte, but salt only melts at very high temperature.

When an electric current is flowing, electrons travel to the cathode, giving it a negative charge, so that it attracts cations. Meanwhile the anode is electropositive and attracts anions. At the electrolyte-electrode interface chemical reactions will occur because of the flow of electrons. At the cathode electrons will be donated, causing reduction reactions, and at the anode electrons will be lost, causing oxidation reactions. So the electrolytic cell is a device for powering redox reactions.

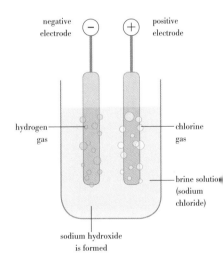

negative electrode

positive electrode

hydrogen gas

chlorine gas

brine solution (sodium chloride)

sodium hydroxide is formed

• Electrolysis of brine solution (aqueous sodium chloride). Reduction and oxidation occur at the interface between the electrodes and the electrolyte, and gas formed as a result bubbles out of solution.

## Electrolysis in Action

Nicholson and Carlisle, the first people to make use of a voltaic pile (see p. 135), stuck electrodes of platinum wire into a dish of water and achieved decomposition of water. As we saw in Chapter 3, even in pure water some $H_2O$ molecules spontaneously dissociate to give $H^+$ and $OH^-$ ions (see p. 97), making it very weakly conductive (so weak that an electrocatalyst such as platinum is needed, and the water may need to be "seeded" with an electrolyte such as an acid). In the decomposition of water, hydrogen cations are attracted to the cathode where they gain electrons and become reduced to hydrogen gas, which bubbles out of solution. At the anode hydroxyl ions are oxidized to give water and oxygen:

$$4OH^-(aq) \longrightarrow O_2(g) + 2H_2O(l) + 4e^-$$

Berzelius used an electrolytic cell to separate salt ions (see pp. 134–135), as when a brine solution is electrolyzed. The chloride anions are attracted to the anode where they are oxidized to chlorine gas:

$$2Cl^-(aq) \longrightarrow Cl_2(g) + 2e^-$$

The $Na^+$ cations migrate to the cathode, but it takes more energy to reduce a sodium ion than a hydrogen one, so hydrogen gas is produced while the sodium effectively forms sodium hydroxide:

$$2Na^+(aq) + 2H_2O(l) \longrightarrow H2(g) + 2NaOH(aq)$$

## Electroplating

The electrode itself can also take part in the electrolytic reactions. For instance, if copper electrodes are immersed in a solution of copper sulfate and a current is passed through it, copper atoms at the anode will be oxidized to copper cations, while copper cations at the cathode will be reduced to copper atoms that are deposited on the surface of the electrode. Eventually the anode will be eaten away. This process can be used to electroplate the cathode—replace the cathode with a metal object and it will be plated with a thin layer of copper. This is how gold and silver plating are done. A similar process is used to extract many metals from their ores. For instance, aluminum is produced by electrolyzing molten aluminum oxide in a carbon tank that acts as the cathode.

---

• **ELECTROLYSIS ON YOUR WRIST**

*A battery or voltaic pile is effectively an electrolysis cell run in reverse, so that electricity is generated rather than consumed. The battery in your wristwatch is an example of a dry cell battery—the zinc outer casing is the anode and in the center of the battery is a steel cathode. The electrolyte is an alkaline paste containing mercury oxide. Most small devices use dry cell batteries; cars use wet cell batteries.*

# 18 Chemical Notation and Electrolysis

## THE PROBLEM:

Berzelius measured the atomic weights of some 43 elements. He introduced a system of using chemical symbols for elements, and numbers to denote their proportions, and developed the concepts of the ion and ionic compounds while working on electrolysis. Will is studying the electrolysis of dilute aqueous sulfuric acid and finds that 36 ml of hydrogen is formed at the negative electrode (cathode). What volume of oxygen will he find at the positive electrode (anode)?

## THE METHOD:

Will needs to write out the equations for the electrolysis of dilute sulfuric acid, $H_2SO_4$, and understand the ions present in solution, and the electron transfers involved, to produce hydrogen and oxygen gases at the electrodes. From this, he can then calculate the ratio of hydrogen to oxygen and thus find the volume of oxygen liberated at the positive anode.

## THE SOLUTION:

Sulfuric acid, $H_2SO_4$(aq), is ionized in aqueous solution and so contains hydrogen ($H^+$), sulfate ($SO_4^{2-}$), and hydroxide ($OH^-$) ions (from the water solvent). Electrolysis due to an electric current passing through the aqueous acid solution causes $H_2SO_4$ to decompose into $O_2$ and $H_2$ gases. Will notes that water itself is also slightly ionized into $H^+$ and $OH^-$ ions:

$$H_2O(l) \longrightarrow H^+(aq) + OH^-(aq)$$

During electrolysis, the positive H⁺ ions are deposited at the negative electrode (cathode) and the negative OH⁻ ions are deposited preferentially (over $SO_4^{2-}$ ions) at the positive electrode (anode). At the respective electrodes, the positive H⁺ ions gain negatively charged electrons (e⁻) to form $H_2$ gas, and the negative OH⁻ ions lose negatively charged electrons to form $O_2$:

$$2 H^+(aq) + 2 e^- \longrightarrow H_2(g)$$
*or*
$$4 H^+(aq) + 4 e^- \longrightarrow 2 H_2(g)$$

(multiplied by 2 to balance number of e⁻)

$$4 OH^-(aq) - 4 e^- \longrightarrow O_2(g) + 2 H_2O(l)$$

Thus, from the equations, it takes an electron gain of two electrons to form each hydrogen molecule ($H_2$) from two H⁺ ions, and the loss of four electrons to make one molecule of oxygen ($O_2$) from four OH⁻ ions. Therefore, the ratio of hydrogen:oxygen formed is 2:1; since the volume of hydrogen was 36 ml, then the volume of oxygen gas formed is:

$$36 / 2 = 18 \text{ ml}$$

- **HOFMANN VOLTAMETER**
  *In 1886, the German chemist August von Hofmann unveiled his eponymous voltameter—a tool for the electrolysis of water. Water, mixed with traces of an ionic compound to increase conductivity, is electrolyzed by a charge applied to platinum electrodes on either side. Gaseous oxygen ($O_2$) collects at the positive anode while gaseous hydrogen collects at the negative cathode, with the gas displacing solution at the top of their respective cylinders.*

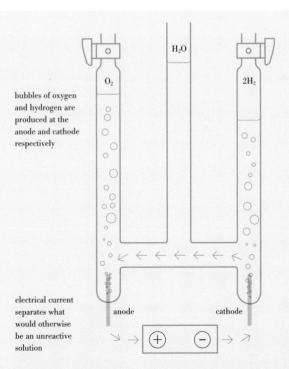

bubbles of oxygen and hydrogen are produced at the anode and cathode respectively

electrical current separates what would otherwise be an unreactive solution

# Humphry Davy

**The great contemporary of Berzelius was the British chemist Humphry Davy, who did more than anyone to raise the profile of his science. Although his contributions to its underlying theory were not as significant as others, his discoveries and inventions made him the most famous chemist of his era, perhaps the best known of all time.**

## The Gas Man Cometh

Humphry Davy (1778–1829) was born in Penzance, a small town at the far southwestern tip of England, to an indigent father and put-upon mother. He was to become the greatest example yet of a man who made good through science, despite which—or more likely, because—he never quite escaped his provincial and penurious background. Like Berzelius, he learned his chemistry from textbooks, devouring Lavoisier's *Traité Elémentaire*, and like Newton, Scheele, and many other notable chemists, he started off as an apothecary's apprentice. In 1798 he was recruited by the radical physician, Thomas Beddoes (1760–1808), for his Pneumatic Institute in Bristol where the latest pneumatic discoveries were to be applied to medicine.

In 1799 Davy published his first papers, which included an insightful attack on Lavoisier's caloric, arguing for a view of heat as motion. He first came to public notice through his experiments with nitrous oxide ($N_2O$), using himself as a guinea pig and thus discovering the psychedelic effects of the gas. He memorably described his "trip": "With the most intense belief and prophetic manner, I exclaimed to Dr. Kinglake, 'Nothing exists but thoughts!—the

universe is composed of impressions, ideas, pleasures and pains!'" His suggestion that nitrous oxide "appears capable of destroying physical pain" was not followed up for 45 years; it is now widely used as an anesthetic. Taking the gas became fashionable and Davy's circle included the Romantic poets Coleridge, Wordsworth, and Southey; he was inclined to poetry himself.

## Brilliant Fragments

In 1800 Davy began working with the new voltaic pile, his analysis demonstrating that the current it generated came from the oxidation of zinc, a finding that secured his election to the Royal Society. In 1801 he moved to London to become the new star of the Royal Institution and began a series of immensely popular lectures. Following Lavoisier's prediction that potash and soda were metal oxides hitherto impossible to decompose, he built the most powerful pile yet constructed (using 250 discs) and with it electrolyzed their molten states and obtained pure potassium and sodium. His cousin Edmund Davy described his reaction: "When he saw the minute globules of potassium burst through the crust of potash, and take fire as they entered the atmosphere, he could not contain his joy—he actually bounded about the room in ecstatic delight." The following year he succeeded in isolating the alkaline earth metals in the same fashion.

While researching acids he decomposed muriatic (hydrochloric) acid and found that it contained not oxygen but chlorine, which he isolated and named (it had been discovered in 1774 by Scheele, who mistakenly thought it was a compound of oxygen). This disproved Lavoisier's oxygen theory and Davy suggested that hydrogen was the agent of acidity. Partly in return for helping British science outshine its French rival, Davy was knighted in 1812. After refusing to patent the safety lamp he had invented (see box), and thus passing up a fortune in royalties, Davy was amply rewarded with fresh honors, including the highest ever bestowed on a scientist: a baronetcy. He became President of the Royal Society but his career petered out in priority disputes, social climbing, and battles over the direction of science, and he spent more time traveling and fishing, dying in Switzerland curiously unfulfilled, in a scientific sense. Berzelius said of him that his work consisted of "brilliant fragments."

## DAVY'S LAMP

The invention that brought him greatest fame was his miner's safety lamp, known as the Davy lamp. In 1815 he was asked to come up with a way of improving mine safety and protecting miners from "mine damp" (build-up of explosive methane). His analysis revealed that if methane is present in the air at a concentration higher than 1 part in 8, the mixture becomes highly flammable and, if it reaches a critical temperature, explosive. This temperature is relatively high, however, and Davy discovered that a metal gauze or mesh can conduct heat away from the flame so quickly that the combustion reaction (the flame) cannot pass through, although light and fumes are able to pass through the apertures.

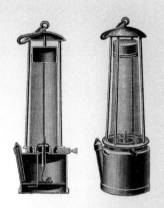

• The Davy lamp could also be used as a gas detector, the flame burning higher or lower or changing color depending on the gases present.

This meant that a cheap and robust safety lamp could be contrived by enclosing the wick in a mesh cylinder or chimney.

# 19 Sodium and Potassium

## THE PROBLEM:

Sodium and potassium were first isolated by Humphry Davy from the electrolysis of "fused" (molten) caustic soda and caustic potash respectively. The metals are both very reactive, and in water they produce "caustic" alkaline solutions (hydroxides) and hydrogen gas in violent reactions. Thus, they are called the alkali metals. Sam is demonstrating the reactions in class and asks her students to calculate how much hydrogen would be liberated at 20°C from reacting 0.1 gram of sodium metal with water.

## THE METHOD:

The students write out a balanced equation for the reaction between sodium and water, knowing that the products are the hydroxide (sodium hydroxide) and hydrogen gas. They then calculate the molar conversion ratio and, from this, can find the volume of $H_2$ gas at 20°C and atmospheric pressure using the molar volume (22.4 liters) or the ideal gas law.

## THE SOLUTION:

The atomic weights of sodium and hydrogen are Na = 23 and H = 1. The balanced equation between sodium metal and water is:

$$2\ Na(s) + 2\ H_2O(l) \longrightarrow 2\ NaOH(aq) + H2(g)$$

| 2 moles | 1 mole |
|---------|--------|
| (46 grams) | (2 grams) |

As you can see, 2 moles (46 grams) of sodium metal reacted with water will give 1 mole (2 grams) of $H_2$ gas. Now that the class has these values, they must rework the equation to calculate the number of moles of $H_2$ produced from 0.1 grams of Na:

0.1 / 23 = 0.0043 moles of Na will produce 0.1 / (2 x 23) = 0.0022 moles of $H_2$ gas.

Remembering that 1 mole of a gas occupies 22.4 L at STP (0°C and 1 atm. pressure), the class can calculate the volume occupied by 0.0022 moles of $H_2$:

$$0.0022 \times 22.4 = 0.049 \text{ L}$$

To find the volume at 20°C, the class must return to the ideal gas law, $PV = nRT$ (see p. 108). As we've seen, P = pressure (in this case 1 atm),

V = volume, $n$ = the number of moles, R = the gas constant (0.082), and T is the temperature in kelvins. As before, the formula must be rearranged so that it can be solved to find the value for V:

$$V = nRT / P$$

Substituting the correct values into this equation, the volume occupied by 0.0022 moles of $H_2$ is:

$$(0.0022 \times 0.082 \times 293) / 1$$

$$= 0.53 \text{ L or } 53.3 \text{ ml}$$

Industrially, sodium hydroxide is still manufactured by the electrolysis of sodium chloride solution (brine). Together with sodium hydroxide and the by-products of hydrogen and chlorine gas, this formed the world-renowned ICI chlor-alkali industry in the UK.

• The reaction of sodium with water produces spectacular results.

# The Periodic Table

The development of inorganic chemistry and

the quest for the elements would reach a climax

with the discovery of the periodic table, a simple

structure that brought together the discoveries of the

chemical revolution into a coherent, unified whole.

This chapter explains the principles that underlie

the periodic table, tells the story of its discovery and

confirmation, and introduces the most important

aspects of subsequent developments in chemistry,

namely nuclear and organic chemistry.

# THE PERIODIC TABLE

The periodic table in its current form runs to 118 elements, although at the upper range of atomic numbers the elements are highly unstable and may have existed for only fractions of a second in a particle accelerator collision chamber. To prevent the table from becoming unmanageably wide, the f-block (lanthanide and actinide series) is usually abstracted and run as a separate block.

This periodic table shows the 118 known elements, while the table below lists the names and atomic weights of the 109 elements for which the names are widely accepted and have been ratified by the International Union of Pure and Applied Chemistry. Color coding indicates the major groups. Note that hydrogen (H) is hard to categorize and in some versions of the table is shown on its own as a block of one.

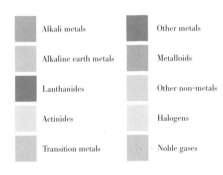

| | | |
|---|---|---|
| Alkali metals | | Other metals |
| Alkaline earth metals | | Metalloids |
| Lanthanides | | Other non-metals |
| Actinides | | Halogens |
| Transition metals | | Noble gases |

| | | | | | | | |
|---|---|---|---|---|---|---|---|
| **Ac** Actinium 227 | **Au** Gold 196.9665 | **Br** Bromine 79.904 | **Cm** Curium 247 | **Ds** Darmstadtium 278 | **Fm** Fermium 257 | **Hf** Hafnium 178.49 | **K** Potassium 39.0983 |
| **Ag** Silver 107.8682 | **B** Boron 10.811 | **C** Carbon 12.0107 | **Cn** Copernicium 285 | **Dy** Dysprosium 162.5 | **Fr** Francium 223 | **Hg** Mercury 200.59 | **Kr** Krypton 83.8 |
| **Al** Aluminum 26.9815 | **Ba** Barium 137.327 | **Ca** Calcium 40.078 | **Co** Cobalt 58.9332 | **Er** Erbium 167.259 | **Ga** Gallium 69.723 | **Ho** Holmium 164.9303 | **La** Lanthanum 138.9055 |
| **Am** Americium 243 | **Be** Beryllium 9.0122 | **Cd** Cadmium 112.411 | **Cr** Chromium 51.9961 | **Es** Einsteinium 252 | **Gd** Gadolinium 157.25 | **Hs** Hassium 277 | **Li** Lithium 6.941 |
| **Ar** Argon 39.948 | **Bh** Bohrium 264 | **Ce** Cerium 140.116 | **Cs** Cesium 132.9055 | **Eu** Europium 151.964 | **Ge** Germanium 72.64 | **I** Iodine 126.9045 | **Lr** Lawrencium 262 |
| **As** Arsenic 74.9216 | **Bi** Bismuth 208.9804 | **Cf** Californium 251 | **Cu** Copper 63.546 | **F** Fluorine 18.9984 | **H** Hydrogen 1.0079 | **In** Indium 114.818 | **Lu** Lutetium 174.967 |
| **At** Astatine 210 | **Bk** Berkelium 247 | **Cl** Chlorine 35.453 | **Db** Dubnium 262 | **Fe** Iron 55.845 | **He** Helium 4.0026 | **Ir** Iridium 192.217 | **Md** Mendelevium 258 |

# The Periodic Table

| 1A | | | | | | | | | | | | | | | | | 8A |
|---|---|---|---|---|---|---|---|---|---|---|---|---|---|---|---|---|---|
| 1 **H** | 2A | | | | | | | | | | | 3A | 4A | 5A | 6A | 7A | 2 **He** |
| 3 **Li** | 4 **Be** | | | | | | | | | | | 5 **B** | 6 **C** | 7 **N** | 8 **O** | 9 **F** | 10 **Ne** |
| 11 **Na** | 12 **Mg** | 3B | 4B | 5B | 6B | 7B | | 8B | | 1B | 2B | 13 **Al** | 14 **Si** | 15 **P** | 16 **S** | 17 **Cl** | 18 **Ar** |
| 19 **K** | 20 **Ca** | 21 **Sc** | 22 **Ti** | 23 **V** | 24 **Cr** | 25 **Mn** | 26 **Fe** | 27 **Co** | 28 **Ni** | 29 **Cu** | 30 **Zn** | 31 **Ga** | 32 **Ge** | 33 **As** | 34 **Se** | 35 **Br** | 36 **Kr** |
| 37 **Rb** | 38 **Sr** | 39 **Y** | 40 **Zr** | 41 **Nb** | 42 **Mo** | 43 **Tc** | 44 **Ru** | 45 **Rh** | 46 **Pd** | 47 **Ag** | 48 **Cd** | 49 **In** | 50 **Sn** | 51 **Sb** | 52 **Te** | 53 **I** | 54 **Xe** |
| 55 **Cs** | 56 **Ba** | 57–71 * Lanthanides | 72 **Hf** | 73 **Ta** | 74 **W** | 75 **Re** | 76 **Os** | 77 **Ir** | 78 **Pt** | 79 **Au** | 80 **Hg** | 81 **Tl** | 82 **Pb** | 83 **Bi** | 84 **Po** | 85 **At** | 86 **Rn** |
| 87 **Fr** | 88 **Ra** | 89–103 ** Actinides | 104 **Rf** | 105 **Db** | 106 **Sg** | 107 **Bh** | 108 **Hs** | 109 **Mt** | 110 **Ds** | 111 **Rg** | 112 **Cn** | 113 **Uut** | 114 **Uuq** | 115 **Uup** | 116 **Uuh** | 117 **Uus** | 118 **Uuo** |

| * Lanthanides | 57 **La** | 58 **Ce** | 59 **Pr** | 60 **Nd** | 61 **Pm** | 62 **Sm** | 63 **Eu** | 64 **Gd** | 65 **Tb** | 66 **Dy** | 67 **Ho** | 68 **Er** | 69 **Tm** | 70 **Yb** | 71 **Lu** |
|---|---|---|---|---|---|---|---|---|---|---|---|---|---|---|---|
| ** Actinides | 89 **Ac** | 90 **Tn** | 91 **Pa** | 92 **U** | 93 **Np** | 94 **Pu** | 95 **Am** | 96 **Cm** | 97 **Bk** | 98 **Cf** | 99 **Es** | 100 **Fm** | 101 **Md** | 102 **No** | 103 **Lr** |

**Mg**
Magnesium
24.305

**Nd**
Neodymium
144.24

**P**
Phosphorus
30.9738

**Pt**
Platinum
195.078

**Rh**
Rhodium
102.9055

**Sg**
Seaborgium
266

**Tc**
Technetium
98

**V**
Vanadium
50.9415

**Mn**
Manganese
54.938

**Ne**
Neon
20.1797

**Pa**
Protactinium
231.0359

**Pu**
Plutonium
244

**Rn**
Radon
222

**Si**
Silicon
28.0855

**Te**
Tellurium
127.6

**W**
Tungsten
183.84

**Mo**
Molybdenum
95.94

**Ni**
Nickel
58.6934

**Pb**
Lead
207.2

**Ra**
Radium
226

**Ru**
Ruthenium
101.07

**Sm**
Samarium
150.36

**Th**
Thorium
232.0381

**Xe**
Xenon
131.293

**Mt**
Meitnerium
268

**No**
Nobelium
259

**Pd**
Palladium
106.42

**Rb**
Rubidium
85.4678

**S**
Sulfur
32.065

**Sn**
Tin
118.71

**Ti**
Titanium
47.867

**Y**
Yttrium
88.9059

**N**
Nitrogen
14.0067

**Np**
Neptunium
237

**Pm**
Promethium
145

**Re**
Rhenium
186.207

**Sb**
Antimony
121.76

**Sr**
Strontium
87.62

**Tl**
Thallium
204.3833

**Yb**
Ytterbium
173.04

**Na**
Sodium
22.9897

**O**
Oxygen
15.9994

**Po**
Polonium
209

**Rf**
Rutherfordium
261

**Sc**
Scandium
44.9559

**Ta**
Tantalum
180.9479

**Tm**
Thulium
168.9342

**Zn**
Zinc
65.39

**Nb**
Niobium
92.9064

**Os**
Osmium
190.23

**Pr**
Praseodymium
140.9077

**Rg**
Roentgenium
283

**Se**
Selenium
78.96

**Tb**
Terbium
158.9253

**U**
Uranium
238.0289

**Zr**
Zirconium
91.224

# PERIODIC TABLE PIONEERS

The rush of elemental discoveries of the early 19th century, and the elucidation of the principles of atomic weight and the law of definite proportions meant that chemistry seemed to be building toward something: a grand synthesis that would unite the microcosmic realm as Newton's laws of gravity had united the macrocosmic realm. Who would achieve this epochal breakthrough?

## The Law of Triads

The synthesis to which chemistry was building was the periodic table, brainchild of the great Russian chemist Dmitry Mendeleyev (see pp. 152–153). But before the discovery of the table there were three important precursors or prototypes, theories that glimpsed a part of the whole that Mendeleyev would unveil so spectacularly in 1869, but which foundered on the incompleteness of chemistry at this time. The first of these periodic table pioneers was German chemist Johann Wolfgang Döbereiner (1780–1849), a lecturer at the University of Jena, amongst whose students was the great writer and philosopher—and dabbler in science—Goethe.

Döbereiner noted that the recently discovered element bromine not only had properties that seemed intermediate between chlorine and iodine, but that it also had an atomic weight that seemed to be midway between the two. Studying the rest of the elements, he spotted two more of these groups of three, which he called "triads": calcium–strontium–barium, and sulfur–selenium–tellurium. In 1829 he announced his "law of triads," but since it appeared to apply to only nine of the 54 elements known at the time it attracted little attention.

## The Telluric Screw

In 1860 the Karlsruhe Conference accepted Avogadro's theories, leading to a revised and far more accurate table of atomic weights for the known elements. French geologist Alexandre-Emile Béguyer de Chancourtois (1820–1886) became the first person to list the elements in order of these revised weights, and he discerned a pattern, which he called the *vis tellurique*, or "telluric screw." The screw was a spiral

De Chancourtois

line drawn down the outside of a cylinder along which were plotted the atomic weights of the elements. Reading down the cylinder gave vertical columns of

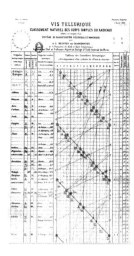

· The vital graph drawn up by de Chancourtois to illustrate his telluric screw concept; published without this explanatory figure, his paper of 1862 fell on stony ground.

elements with the same properties, according to a period of 16 units of atomic weight—in other words, a pattern where the properties repeated periodically every 16 amu.

Unfortunately, when de Chancourtois published his paper explaining the screw in 1862, the journal failed to include the explanatory diagram, making it almost impossible for the reader to visualize. It did not help that he had couched his theories in the language of geology and also veered into the esoteric territory of numerology. Inevitably, few people took any notice.

## The Law of Octaves

Just two years later English chemist John Newlands (1837–1898) listed the elements in ascending order of atomic weight, in vertical lines of seven (the noble gases, which would have brought this up to eight, as in the modern periodic table, had not yet been isolated), and discovered that this pattern gave rows of elements with similar properties. Much taken with musical theory, and perhaps with Pythagorean ideas of mystical numbers and the music of the spheres, he interpreted this via the concept of octaves: "The eighth element starting from a given one is a kind of repetition of the first, like the eighth note in an octave of music." He declared this to be the "law of octaves." Newlands reported his theory in a paper to the Chemical Society in 1865, but there were many holes in his scheme, particularly at higher atomic weights, where any apparent pattern of correspondences broke down. (Like Döbereiner and de Chancourtois before him, Newlands had to contend with gaps in contemporary knowledge, including undiscovered elements and incorrectly calculated atomic weights. Mendeleyev's genius was to overcome these.) Newlands was met with ridicule, with one commentator suggesting that he might as well arrange the elements alphabetically. After Mendeleyev's system was published he claimed priority, although his scheme had lacked precisely the innovations that made the Russian's so brilliant. Eventually the Royal Society awarded him the Davy Medal in 1887, but he was never elected a member.

# Dimitry Mendeleyev

**Hailed as the greatest chemical mind since Lavoisier, Dimitry Mendeleyev did important work for industrial and agricultural chemistry, helped regulate Russia's weights and measures, and authored a standard textbook. His most enduring achievement, however, was his periodic law or, as it is better known, the periodic table, which ranks alongside the achievements of Newton and Darwin.**

## The Fundamental Theme

Born to an impressive mother who ran a glass factory in Siberia to support her family when her husband died, Dimitry Mendeleyev (1834–1907) was the youngest in a large family. A brilliant student, he overcame illness to win a scholarship to study in Germany with Robert Bunsen (of burner fame, see pp. 164–165). In 1861 he returned to Russia to take up a position at St. Petersburg University. Like many others around this period he was preoccupied with uncovering what he called, "the philosophical principles of our science which form its fundamental theme."

Working on a new textbook in 1869, Mendeleyev was prompted to consider the issue of whether the elements could be arranged according to some system or law. One of the few people aware of the work of de Chancourtois, Mendeleyev began to play with the order of the elements for himself, noting that the halogens and the oxygen and nitrogen groups of elements could be arranged in a table of ascending atomic weights. Seeking a larger pattern that included all the other elements, he wrote the name and atomic weight of each one on a card and arranged them in vertical lines. After working on the problem fruitlessly for three days he fell asleep and had a celebrated dream: "I saw in a dream a table where all the elements fell into place as required. Awakening, I immediately wrote it down on a piece of paper." His dream table clearly showed that, if arranged according to atomic weight, the elements followed a periodic law (see pp. 154–155).

## A Suggested System

His historic paper "A Suggested System of the Elements" showed a table in which the elements were ordered in columns of descending atomic weight, arranged such that

• With his trademark long hair and beard, Mendeleyev cut an imposing figure. He first came to international attention with his landmark 1870 textbook *Principles of Chemistry*, which was translated into many languages.

each row contained elements with similar properties. What was revolutionary and daring about his scheme was its refusal to adhere to the constraints that had hamstrung previous efforts. Where necessary he put some elements out of order (putting question marks next to their atomic weights) and left gaps where there was no element that fit the pattern. Although breaking one of the cardinal rules of science—namely that theories should be made to fit the evidence, rather than vice versa—this was the precisely the intuitive leap necessary to solve an intractable problem. In effect Mendeleyev was predicting that his theory was right, and that where chemistry disagreed with him, it was science that was wrong.

The true test of a scientific theory is that it makes testable predictions (see p. 83), and Mendeleyev's periodic law did just this. Not only was he able to predict which atomic weights had probably been incorrectly determined, he was even able to predict the existence of hitherto unknown elements, including their likely

· This monument to the periodic table can be found at the Slovak University of Technology in Bratislava, Slovakia. Dimitry Mendeleyev's distinctive portrait occupies center stage.

atomic weights and even their properties. These unknown elements included one between aluminum and indium, which he named eka-aluminum and predicted would have an atomic weight of 68, and an element of atomic weight 70 between silicon and tin that he named eka-silicon (eka was Sanskrit for "one"—as in "aluminum, or silicon, plus one").

Confirming his confidence that the table was accurate was the correspondence between each of his horizontal rows—or families—and the valence (see p. 27) of the elements it contained. Reading vertically along the table, the valencies went from 1 on the lithium row up to 4 on the carbon row and back down to 1, giving a pattern of 1, 2, 3, 4, 3, 2, 1—a periodic rise and fall. Here was the periodic law he had been looking for. Although there were inconsistencies, he was confident enough to overlook these: "Although I have had my doubts about some obscure points, yet I have never doubted the universality of this law, because it could not possibly be the result of chance."

### ОПЫТЪ СИСТЕМЫ ЭЛЕМЕНТОВЪ.

ОСНОВАННОЙ НА ИХЪ АТОМНОМЪ ВѢСѢ И ХИМИЧЕСКОМЪ СХОДСТВѢ.

|  |  |  |  |
|---|---|---|---|
|  | Ti = 50 | Zr = 90 | ? = 180. |
|  | V = 51 | Nb = 94 | Ta = 182. |
|  | Cr = 52 | Mo = 96 | W = 186. |
|  | Mn = 55 | Rh = 104,4 | Pt = 197,4. |
|  | Fe = 56 | Ru = 104,4 | Ir = 198. |
|  | Ni = Co = 59 | Pl = 106,6 | O = 199. |
| H = 1 | Cu = 63,4 | Ag = 108 | Hg = 200. |
| Be = 9,4 | Mg = 24 | Zn = 65,2 | Cd = 112 |
| B = 11 | Al = 27,4 | ? = 68 | Ur = 116 | Au = 197? |
| C = 12 | Si = 28 | ? = 70 | Sn = 118 |
| N = 14 | P = 31 | As = 75 | Sb = 122 | Bi = 210? |
| O = 16 | S = 32 | Se = 79,4 | Te = 128? |
| F = 19 | Cl = 35,5 | Br = 80 | I = 127 |
| Li = 7 | Na = 23 | K = 39 | Rb = 85,4 | Cs = 133 | Tl = 204. |
|  | Ca = 40 | Sr = 87,6 | Ba = 137 | Pb = 207. |
|  | ? = 45 | Ce = 92 |
|  | ?Er = 56 | La = 94 |
|  | ?Yt = 60 | Di = 95 |
|  | ?In = 75,6 | Th = 118? |

Д. Менделѣевъ

· The original Russian version of Mendeleyev's periodic table, which is oriented perpendicular to modern versions, so that the periods run vertically. Note the question marks next to elements that he predicted but which had not yet been discovered.

# PERIODIC LAW

The periodic law that Mendeleyev had uncovered, albeit somewhat refined in subsequent years, is the key to inorganic chemistry. With this law chemists can make sense of both the big picture and the fine details of their field, grouping the elements into families with similar physical and chemical properties, and predicting how they will interact and even which ones have yet to be discovered.

## Ordering the Elements

Mendeleyev, like the periodic pioneers before him, ordered the elements in his system according to their atomic weight. At the time, the concept of subatomic particles was pure speculation and there was no way to know of the existence of protons, let alone count them. This caused problems for the new periodic system because atomic weight only correlates with chemical properties in as far as it correlates to atomic number, for it is the number of electrons an element possesses that determines its chemistry (see pp. 26–27), and this in turn is decided by the number of protons (its atomic number). Hence the modern periodic table is ordered by atomic number, not weight, resolving some inconsistencies in Mendeleyev's original.

In the periodic table, the elements are sorted into seven rows or periods, with the atomic number increasing as you move along the row from left to right. This arranges the elements into columns that are called families, because in each column the elements share a "family resemblance," with similarities in physical and chemical properties. Take a look at the periodic table on page 149 and you will see that the first four periods consist of 2, 8, 8, and 18 elements respectively. What does this pattern tell us about where the periodicity comes from? These numbers correspond to the size of the valence shell of each period. The first period consists of hydrogen (atomic number 1) and helium (atomic number 2), atoms with only the innermost electron shell, which can only contain two electrons, as their valence shell. The next electron shell available can have up to eight electrons, as can the third, while the fourth can hold up to 18. Actually the picture is slightly more complicated than this, as the shells are subdivided into orbitals designated by the letters $s$, $p$, $d$ and $f$. Periods six and seven of the periodic table have 32 elements each, and at such high atomic numbers the electron configurations become extremely complex, with $f$ orbitals that can hold up to 14 electrons.

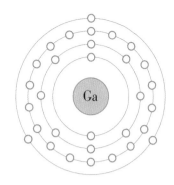

• The electron configuration of gallium (Ga), showing how it has three electrons in its outer orbital. This gives it a valence of 3, and puts it in the p-block because it is just beginning to fill its outer, 4p-orbital.

## Predictive Power

As Mendeleyev proved when he first formulated it, periodic law is a powerful tool that allows chemists to predict the potential existence of elements not yet discovered—in fact not yet created, because they may only come into existence if produced artificially in high-energy nuclear physics experiments. At first, Mendeleyev's prediction of the existence of eka-aluminum and eka-silicon looked like hubris, for several years went by without any such announcement. But French chemist Paul Lecoq de Boisbaudran (1838–1912) knew of Mendeleyev's prediction and resolved to find the missing element. Lecoq knew that eka-aluminum would have an atomic weight of 68 or thereabouts, and concluded that the best place to look for it would be in an ore of zinc, which has an atomic weight of around 65. After much laboring he succeeded in 1875 in identifying by

spectroscopy (see p. 164) a new element, which he named gallium, and which had an atomic weight of 69.72. In fact zinc and gallium are atomic numbers 30 and 31, which is why they are neighbors in the table. When Lecoq analyzed the new metal he found that its density was 4.9 $g/cm^3$; this did not match Mendeleyev's prediction of 6.0 $g/cm^3$. The Russian was so confident in his own theory that he told Lecoq that he was the one who had made the mistake, probably because his sample was contaminated. The Frenchman duly went back and remeasured the density of a purer sample, obtaining a value of 5.9 $g/cm^3$.

Further discoveries offered yet more confirmation that Mendeleyev was right. In 1879 the element scandium was discovered, matching his prediction of eka-boron, and in 1886 his eka-silicon was found, and named germanium.

• A sample of gallium, a soft, silvery metal that melts in the hand. Its compounds are extremely valuable as semiconductors and used in microwave and light-emitting diode technologies.

# 20 Electrons, Shells, and Valency

## THE PROBLEM:

J. J. Thomson discovered the electron and the existence of isotopes. Electrons were proved to be negative particles and later shown by Rutherford and Bohr to surround the nucleus in energy levels or shells at certain distances (represented as orbits) from the nucleus. The electron arrangement within the shells is called the electronic configuration. Using the periodic table, chemistry student Abe wants to construct the basic electronic configurations for sodium, chlorine, magnesium, and oxygen.

## THE METHOD:

In the periodic table, the elements are placed in order of increasing atomic number (the number of positively charged protons in the nucleus) and so the nucleus is surrounded by an equal number of negatively charged electrons for the atom to be electrically neutral. Each shell can only hold up to a maximum number of electrons which is 2, 8, and 18, respectively, for the first three shells going outward from the nucleus. The number of (valence) electrons in the outermost shell also gives the element its group number and allows its valency to be deduced.

## THE SOLUTION:

The element sodium has an atomic number of 11 and so has 11 electrons surrounding its atomic nucleus. Applying the rules above, the arrangement of 11 electrons within the three shells would be 2, 8, and 1, and the electronic configuration represented as [2, 8, 1] with the one outermost electron (in the 3rd shell) placing sodium in group 1. Similarly, magnesium, with atomic number 12, would have an electronic arrangement of [2, 8, 2] and be in group 2. On the other hand, chlorine has an atomic number of 17 giving a configuration of [2, 8, 7] and is found in group 7. Finally, oxygen, with atomic number 8, will have a configuration of [2, 6], placing it in group 6. Apart from the first shell, which is full with two electrons, the other shells are stable with a maximum of eight electrons although the third can contain a maximum of 18.

The valency of an element can be generally defined as the number of electrons with which a given atom generally bonds, or the number of bonds an atom forms. So, in the case of sodium and magnesium, with respectively one and two outermost (valence) electrons, their valencies are 1 and 2, respectively, as losing one and two electrons would give a stable outer configuration of eight electrons (stable octet). On the other hand, chlorine and oxygen would have to gain one and two electrons respectively to gain a stable octet and thus will have respective valencies of 1 and 2. A stable octet can be obtained by the transfer (loss and gain) of electrons between atoms in ionic bonding, and by the sharing of electrons in covalent bonds in covalent compounds.

• The electron configuration of sodium (Na), showing the single valence electron in its outer shell.

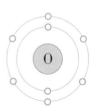

• The electron configuration of chlorine (Cl), showing the seven electrons in its valence shell, which, following the octet rule, give it a valency of 1.

• The electron configuration of magnesium (Mg), which has one more electron than sodium and hence a valency of 2.

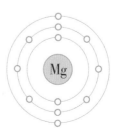

• The electron configuration of oxygen (O), which lacks two electrons to fill its outer orbital and hence has a valency of 2.

# THE PERIODIC TABLE

The periodic table arranges elements into groups and families, making their categorization and description much simpler. Understanding these family relationships is essential for any student of chemistry. Grasping the underlying rules that determine the family characteristics immediately reduces the apparent complexity of chemistry, helping to make sense of the confusion of names and terms.

## Metals, Non-metals, and Metalloids

There are various ways of dividing up the elements in the table. One is to split them into three broad categories, metals, non-metals, and metalloids. The metals are all the elements to the left of a stepped line drawn down the table, starting at element 5, boron (B), and extending down to polonium (element 84, Po), with the exception of germanium (Ge) and antimony (Sb). The non-metals (together with hydrogen) are to the right of the line, while the elements bordering the line are the metalloids. Figure 1 shows the metalloids extracted from the table.

Metals have physical properties that are familiar from everyday life. They are almost all solids [the only one that is liquid at room temperature is mercury (Hg), although cesium (Cs) and gallium (Ga) both melt at 86°F (30°C)]. Most are hard, dense and shiny, and give a characteristic "ping" if struck. They are ductile and malleable (they can be drawn into thin wires or hammered into flat sheets). Chemists classify a metal by conductivity—metals are good conductors of electricity and heat. In chemical reactions they generally lose, or donate, electrons.

Non-metals include several gases and liquids; when solid they tend to be brittle. They are poor conductors and in chemical reactions they tend to gain electrons. Metalloids, also known as semimetals, combine characteristics of the other two groups, including conductivity. As a result they are

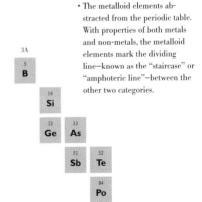

• The metalloid elements abstracted from the periodic table. With properties of both metals and non-metals, the metalloid elements mark the dividing line—known as the "staircase" or "amphoteric line"—between the other two categories.

## SUPER-HEAVY ELEMENTS

At very high atomic numbers the nucleus of the atom becomes very large and unstable, and the elements become radioactive (see pp. 168–169), decaying (breaking down) to give elements with lower atomic numbers. This means that the highest naturally occurring elements are uranium and plutonium; if elements heavier than this should happen to form, perhaps in a nuclear reaction, they will almost instantaneously decay. However, atom-smashing technology has allowed scientists to create artificially many of the super-heavy elements predicted by the periodic table. The most recently created element, at the time of writing, is ununseptium (Uus), element 117. In January 2010 a US-Russian team claimed to have created six atoms of the element by smashing together atoms of calcium (number 20) and berkelium (number 97). The heaviest atom ever created is ununoctium (Uuo, 118), of which only four atoms with a half-life of just milliseconds have ever been observed. The names for these super-heavy atoms are "placeholders," created according to a system set out by the International Union of Pure and Applied Chemistry (IUPAC), until permanent names and symbols can be agreed. The system uses Latin and Greek roots and suffixes; for example, element 119, should it ever be created, will be called ununennium: un (1) + un (1) + enn (9) + ium (standard element suffix). Agreeing permanent names tends to be a political process fraught with controversy. For instance, ununtrium (Uut, 113) has had two names proposed—japonium and rikenium; it is up to IUPAC to decide between them.

semiconductors, which makes them useful in electronics and thus economically valuable.

Most of the elements in the periodic table form oxides; how these behave in water is generally determined by whether the element is metallic or non-metallic. Metals form basic oxides, which is to say that they react with water to form alkaline solutions. Nonmetals form acidic oxides, which react with water to give acid solutions. Some elements, such as aluminum, form amphoteric oxides, which can go either way.

### Family Ties

Another way of dividing up the periodic table is into its vertical columns, which form families. These are numbered, either from 1 to 18, or, more traditionally, with Roman numerals and letters. In the latter system periods 1 and 2 and the elements below them in these columns are families IA-VIIIA. The columns numbered 3–12 in the modern system are "B" families in the old system, with columns 8–10 grouped together as a single family, VIIIB, and, confusingly, columns 11 and 12 labeled IB and IIB.

It is worth looking in greater detail at some notable families: the alkali metals, IA (modern system: 1); the alkaline earth metals, IIA (modern system: 2); the halogens, VIIA (modern system: 17); and the noble gases, VIIIA (modern system: 18). (Hydrogen is placed on the table as if it were a member of IA, but this is only because of its atomic number. In practice, hydrogen is anomalous and is in a family of its own.)

• The six elements that comprise the group of alkali metals.

| Li | K | Cs |
|---|---|---|
| Lithium | Potassium | Cesium |
| 6.941 | 39.0983 | 132.9055 |
| Na | Rb | Fr |
| Sodium | Rubidium | Francium |
| 22.9897 | 85.4678 | 223 |

IA
| 3 Li |
| 11 Na |
| 19 K |
| 37 Rb |
| 55 Cs |
| 87 Fr |

• The six elements that constitute the alkaline earth metals.

| Be | Ca | Ba |
|---|---|---|
| Beryllium | Calcium | Barium |
| 9.0122 | 40.078 | 137.327 |
| Mg | Sr | Ra |
| Magnesium | Strontium | Radium |
| 24.305 | 87.62 | 226 |

2A
| 4 Be |
| 12 Mg |
| 20 Ca |
| 38 Sr |
| 56 Ba |
| 88 Ra |

The alkali metals are highly reactive and soft enough to cut with a knife, and all have one outermost or valence electron that they lose to form ions with a single positive charge and thus assume an oxidation state of I. The alkaline earth metals are also generally very reactive; like the IA metals they are found in nature as ionic salts, but because they each have two valence electrons they adopt oxidation state II. The halogens get their name from their tendency to react with metals to form salts (*halx* in Greek). They each have seven valence electrons so they tend to be strong oxidants, gaining one electron to form ions with a

## • THE LIMITS OF THE PERIODIC TABLE

*Given that new elements are still being found, how many periods could the periodic table have? Once the f orbital is filled, the next orbital—should it prove to exist—will be called the g orbital, so if the eighth and ninth periods do exist they would contain a g block, which would start at element 121, unbiunium. If there is a theoretical upper limit on the size of an atom it is probably linked to the maximum number of neutrons that will stay in a nucleus before it decays. Certain 'magic numbers' of neutrons and protons in combination may give elements constituting an 'island of stability', in terms of nuclear decay, at the far reaches of the Table, possibly with atomic numbers as high as 164.*

| 9 | |
|---|---|
| **F** | |

| 17 | |
|---|---|
| **Cl** | |

| 35 | |
|---|---|
| **Br** | |

| 53 | |
|---|---|
| **I** | |

| 85 | |
|---|---|
| **At** | |

• The five elements that make up the group of halogens.

| **F** | **Cl** | **Br** |
|---|---|---|
| Fluorine | Chlorine | Bromine |
| 18.998 | 35.453 | 79.904 |

| **I** | **At** |
|---|---|
| Iodine | Astatine |
| 126.905 | 210 |

single negative charge. All three of these families display trends typical of periodic families—their characteristic family properties tend to get weaker as you go down the column, although the first member of each family is often slightly anomalous (for example, the chemistry of lithium is different from the other alkali metals).

The noble gases were not known at the time Mendeleyev constructed his table, and when they were discovered his initial reaction was one of dismay that an entirely new class of elements would destroy his theory. In fact they proved to be the final piece of the puzzle, slotting in neatly at the end of the table. They have complete valence shells of eight electrons, making them extremely unreactive. For these A families, their Roman numerals predict the size of their valence shells and thus their chemical properties.

| **He** | **Ne** | **Ar** |
|---|---|---|
| Helium | Neon | Argon |
| 4.003 | 20.179 | 39.948 |

| **Kr** | **Xe** | **Rn** |
|---|---|---|
| Krypton | Xenon | Radon |
| 83.798 | 131.293 | 222 |

• The six naturally occurring noble gases all possess very low reactivity.

| 2 | |
|---|---|
| **He** | |

| 10 | |
|---|---|
| **Ne** | |

| 18 | |
|---|---|
| **Ar** | |

| 36 | |
|---|---|
| **Kr** | |

| 54 | |
|---|---|
| **Xe** | |

| 86 | |
|---|---|
| **Rn** | |

## Letter Blocks

A third way of dividing up the table is by electron orbitals. While each period corresponds to electron shells successively more distant from the nucleus, as mentioned on page 154, these electron shells are subdivided into $s$, $p$, $d$ and $f$ orbitals (containing up to 2, 6, 10 and 14 electrons respectively). As atomic number increases across each period, so elements begin to fill each of these in turn, and the table can be split into blocks corresponding to these orbitals. The $s$ block on the left consists of families IA and IIA, while the $p$ block on the right consists of families IIIA–VIIIA—moving from left to right along these periods the $p$ orbital is progressively filled. Between these is the $d$ block, containing the B families known as the transition metals, where, as you move across the period, electrons progressively fill the $d$ orbital. Because the $d$ orbital can hold up to 10 electrons, the $d$ block is 10 elements across. When you get to period 6 and 7 the $f$ orbital kicks in, but to save space by cutting down on the width of the periodic table, the $f$ block is usually shown pulled out of the table as a separate block. It consists of elements known as the lanthanides or rare earths (on period 6), and the actinides (on period 7), which are all radioactive elements. Because the $f$ orbital can hold up to 14 electrons, the $f$ block is 14 elements across.

# The Periodic Table

## THE PROBLEM:

Mendeleyev created the periodic table of the chemical elements to show recurring (periodic) trends in properties of the then known elements, listed in order of increasing atomic weight. As new elements were discovered, the layout has been changed and the modern table now contains 118 elements listed in order of their increasing atomic number. Iron is an essential element to sustaining life; Cath wonders if it is possible to use the periodic table (see p. 149) to identify the number of protons and electrons in the iron (Fe) atom and in $Fe^{2+}$ and $Fe^{3+}$ ions.

## THE METHOD:

The periodic table shows the elements in order of increasing atomic number (the number of protons in the nucleus of the atom), starting from hydrogen (atomic number 1) to ununoctium (atomic number 118), with their atomic weights based on a weighted mean of their isotopes. An isotope of an individual element has the same atomic number but a different mass number (the number of protons plus neutrons in the nucleus of the atom). Since each atom of an element is electrically neutral, then the number of positively charged protons in the nucleus (atomic number) must equal the number of negatively charged electrons ($e^-$) outside the nucleus (see pp. 132). A positive ion is thus formed by loss of one or more electrons.

## THE SOLUTION:

Using the periodic table, the atomic number of iron (Fe) is 26, which means that the neutral iron atom (Fe) has 26 positively charged protons within the nucleus, and therefore 26 negatively charged electrons outside the nucleus. Thus, to obtain the $Fe^{2+}$ and $Fe^{3+}$ ions, we need to lose respectively two and three electrons:

$$Fe - 2\ e^- \longrightarrow Fe^{2+}$$
$$Fe - 3\ e^- \longrightarrow Fe^{3+}$$

Therefore, in the $Fe^{2+}$ ion, there are 26 protons and $(26-2) = 24$ electrons. Similarly, in the $Fe^{3+}$ ion, there are 26 protons and $(26-3) = 23$ electrons, which give the relevant ionic charge.

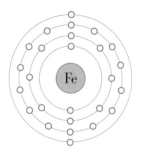

| 26 | **Fe** |
|---|---|
| **Fe** | Iron |
| | 55.845 |

• The electron configuration for iron (Fe), showing that its apparent valence shell is the 4s orbital. In practice the sub-shells are more complex than shown here, and iron readily assumes two different oxidation states. The periodic table entry shows its atomic number and atomic weight; the latter is not an integer because more than one isotope of iron is found in nature.

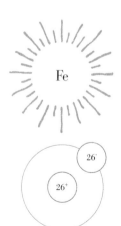

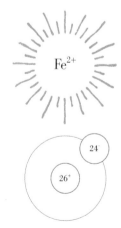

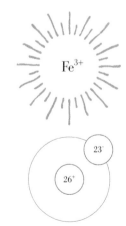

# SPECTROSCOPY

As periodic law reveals, chemistry and physics are intimately interrelated. The voltaic pile, a technology born of chemistry, had ushered in the 19th century and the beginnings of the modern age by opening new worlds in chemistry and physics; now a new technology born of the marriage between chemistry and physics would extend the grasp of both to literal new worlds.

## Fingerprints of Light

German optician Joseph von Fraunhofer (1787–1826) was the first to notice that the spectra (range of wavelengths) of light from a flame, when viewed through optical glass, was characterized by distinct lines of brightness. Directing his glasses at the Sun, he discovered that the continuous spectrum of its light was broken up by a number of dark lines, which he labeled with letters. These spectral lines remained something of a mystery; in 1826 William Henry Fox Talbot suggested that they might be used for chemical analysis, but nothing came of this until the 1850s.

It was at this time that Robert Bunsen (1811–1899), professor of chemistry at Heidelberg (and teacher of Mendeleyev among others), noted that burning some elements produced flames with characteristic colors. He approached Heidelberg's professor of physics, Gustav Kirchhoff (1824–1887) to help him analyze the problem, and in 1859, using a prism to separate the different wavelengths by refraction, they were able to obtain spectra from different elements, and prove that each one emitted a unique and stable spectrum that could be used to identify it—in other words, a spectral fingerprint.

Known as spectrochemical analysis, in 1860–61 this technique allowed Bunsen to prove the existence of two previously unknown elements found in mineral water in trace amounts, including one that produced deep red spectral lines and was therefore named rubidium (from the Latin for "deep red"). Also in 1861, William Crookes discovered thallium using the same process. Meanwhile Kirchhoff

• An etching from 1900 depicting Joseph von Fraunhofer as he demonstrates the spectroscope to an admiring audience.

• Ubiquitous in school laboratories and the object of much misuse, the Bunsen burner is a small gas burner that allows the user to control the mixture of air and gas (and thus the size and intensity of the flame) with a simple valve, while providing a stable and clean-burning flame.

ingeniously applied the technique in reverse to analyze sunlight, and he and Bunsen were able to show that von Fraunhofer's dark "D lines" corresponded exactly with the bright yellow lines produced by sodium. They deduced that sodium must be present in the atmosphere of the Sun, and rather than emitting yellow light as when it was burned in a flame, it was absorbing it and blocking those portions of the spectrum from reaching Earth. Science had now made it possible to analyze the elements of the stars.

## Quantum Leap

Emission spectroscopy, the analysis of spectra emitted by a substance, depends on the rules governing the energy levels, or states, of electrons orbiting an atom.

When an electron absorbs a packet of light energy known as a quantum it can jump from its resting or ground state to an excited state. When it falls back it releases the quantum, emitting it as light. The number of quanta needed depends on the orbitals in question; this in turn determines the energy and therefore wavelength of the light emitted when the electron falls back from its excited state to its ground state. Each element has a unique configuration of electrons, and therefore a unique absorption and emission spectrum, meaning that the spectrum can be used to read the electron configuration and, by extension, the atomic number of the element.

---

## • LIGHT INSTRUMENTS

*The basic instrument of spectroscopy is the spectroscope, which consists of a slit to allow only a beam of light through, a collimator (a device for narrowing a beam of light so that its rays are parallel), a prism or grating to separate wavelengths by refraction or diffraction, and a telescope or microscope objective lens for the user to look through. Adding a camera or some other recording device to the instrument makes it into a spectrograph, while adding a calibrated scale for measuring the spectra makes it a spectrometer.*

• Simplified representation of a spectrograph, with a beam from a collimator (entering from the right) split into a spectrum by a prism and focused onto a photographic plate.

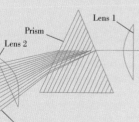

Lens 1

Prism

Lens 2

Photographic plate

Spectrum

### THE PROBLEM:

Using his mass spectrograph to separate positively charged isotope (M⁺) ions, British scientist Francis William Aston identified 212 out of 287 naturally occurring isotopes. Aston's discovery has enabled paleontologists to carbon date fossilized remains by analysing the presence of the three main isotopes of carbon: carbon-12, carbon-13, and carbon-14 (the technique involves comparing the differences in the ratios of these isotopes). To identify each isotope, the number of protons and neutrons present in the nucleus must be determined. How is this done and what is the mass of the three isotopes of carbon?

### THE METHOD:

The nuclear symbol indicates the composition of the nucleus, shown with the atomic number (number of protons) as subscript, and the mass number (protons and neutrons) as superscript, both to the left of the element symbol as standard convention. From both the atomic number (protons) and the mass number (mass), the number of neutrons (by difference) can be found.

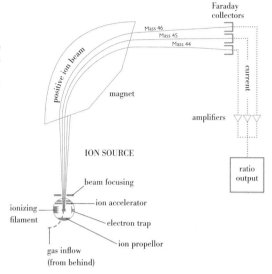

• A simplified diagram illustrating the components of a mass spectrometer. First developed in the early 20th century, it determines the mass of a molecular substance by vaporizing and ionizing a substance, before calculating the properties of the ions as they move through electro-magnetic fields.

## THE SOLUTION:

In an atom of carbon-12 ($^{12}$C), the number of protons (atomic number) is 6 and the mass number (number of protons and neutrons, its mass) is 12, which means that there are (12 − 6) = 6 neutrons in the nucleus.

In an atom of carbon-13 ($^{13}$C), the atomic number is 6 and the mass number is 13, which means that there are (13 − 6) = 7 neutrons in the nucleus.

In an atom of carbon-14 ($^{14}$C), the atomic number is 6 and the mass number is 14, which means that there are (14 − 6) = 8 neutrons in the nucleus. The isotopes are thus different in mass due to the different number of neutrons in the nucleus of the atom. In an electrostatic and magnetic field, the lighter isotope ions are deflected the most and the heavier ions are deflected

the least leading to separation of the isotopes by mass in spectrometry.

Mass spectrometry is a powerful technique and, apart from identifying isotopes, is used in organic chemistry to determine the molecular weight of a pure organic compound, and also to identify the components in a mixture by linking to a gas chromatograph (GC). The "tandem" technique of gas chromatography-mass spectrometry (GC-MS) uses the mass spectrum as a chemical fingerprint to identify the compounds separated by the gas chromatograph. GC-MS was used on the Cassini-Huygens space mission to explore the surface of Saturn and its moon, Titan, in the search for extraterrestrial life.

# RADIOACTIVITY AND ISOTOPES

Normal chemistry deals almost exclusively with the action and behavior of electrons; atomic nuclei hardly get a look in, as for the most part they are spectators to the phenomena of covalent and ionic bonding, electrochemistr and the like. When the nucleus does get involved it is termed nuclear chemistry, which deals with radioactivity, isotopes, and nuclear reactions.

## Isotopes

As we have seen, isotopes are atoms with the same atomic number but with more or fewer neutrons and therefore different atomic weight. For instance, carbon-14 and carbon-12 both have 6 protons and therefore both have the same atomic number (Z). However, carbon-14 has eight neutrons in its nucleus whereas carbon-12 has only six, giving them atomic masses (A) of 14 and 12, respectively. Because both isotopes have the same number of protons, they also have the same number of electrons and thus the same chemistry.

Another example is uranium-238 and uranium-235; for both isotopes $Z = 92$, but the former has 146 neutrons and the latter only 143. The number of neutrons in an isotope is A–Z. In scientific notation, an isotope is indicated by giving the atomic mass as a superscript in front of the element symbol:

$$^{14}C \text{ and } ^{12}C; \, ^{238}U \text{ and } ^{235}U$$

The atomic weight of an element is the average of the atomic masses of all its isotopes, weighted to take into account their frequency. The vast majority of carbon atoms, for instance, are $^{12}C$, so the atomic weight of carbon is very close to 12 (A=12.0115). Of the 83 elements which occur naturally in significant quantities on Earth, 20 are found as a single isotope (mononuclidic), and the others as mixtures of up to 10 isotopes.

## Radioactive Decay

Radioactivity is the decomposition or decay of an unstable nucleus, involving the loss and/or transformation of subatomic particles with release of energy. Stability of the nucleus is determined by the proton:neutron (P:N) ratio. If an isotope has too few or too many neutrons it will be unstable; the stable ratio depends on the atomic number. Nuclei also become unstable when they get above a certain size; all the elements with an atomic number of 84 or more are unstable and therefore radioactive. Radioactive decay takes places when the nucleus "wants" to achieve a more stable P:N ratio. This can involve three types of radiation: alpha particles, beta particles, and gamma rays.

Alpha particles consist of two protons and two neutrons; they are effectively helium cations (helium atoms stripped of their electrons). When an atom emits an alpha particle its atomic mass falls by 4 amu and its atomic number by 2. Alpha particle emission is typical of heavy elements, like uranium. Decay is a form of nuclear reaction and can be described in an equation similar to that used in chemical reactions. For example:

$$^{238}_{92}\text{U} \longrightarrow {}^{234}_{90}\text{Th} + {}^{4}_{2}\text{He}$$

A beta particle is an electron that is emitted from the nucleus when a neutron decays into a proton plus an electron. The electron shoots out of the nucleus, leaving the proton behind. This means the atomic mass does not change but the atomic number increases by 1. For example, the hydrogen isotope known as tritium has two neutrons and one proton. This is an unstable P:N ratio, so one of the neutrons decays to emit a beta particle, becoming a proton in the process. This changes the atomic number to 2, converting the hydrogen atom into an isotope of helium:

$$^{3}_{1}\text{H} \longrightarrow {}^{3}_{2}\text{He} + {}^{0}_{-1}\text{e}$$

Note that the beta particle, even though it is just an electron, is written with the specific notation:

$$^{0}_{-1}\text{e}$$

## DREAM ACHIEVED

Radioactive decay is a form of transmutation, because it accomplishes what the alchemists always dreamed of: changing one element into another. Ironically, naturally occurring transmutation is more likely to change a valuable element into a base metal than vice versa—uranium, for instance, decays in a series of steps until it becomes lead. Artificial transmutation is possible with atom smashers, which can add protons and neutrons to elements.

This is to allow the equation to be balanced. Just as with a chemical equation, the numbers on both sides must match up—in the case of a nuclear reaction, the atomic mass and numbers must match up. In this case $(3 = 3 + 0)$ and $(1 = 2 + -1)$.

Gamma radiation is a form of electromagnetic energy. Sometimes a nucleus that has undergone alpha or beta decay is left in an excited state and drops down to a lower energy state by emitting a very high-energy (and therefore high-frequency) photon, known as a gamma ray. Gamma rays are close to X-rays on the electro-magnetic spectrum.

# Marie and Pierre Curie

**The key figures in the elucidation of radioactivity were a husband and wife team, Marie and Pierre Curie. Marie has become particularly celebrated; the first person to win two Nobel prizes, she overcame hardship and prejudice to blaze a trail for women scientists. The Curies' pioneering research helped reveal the chemistry of radioactivity.**

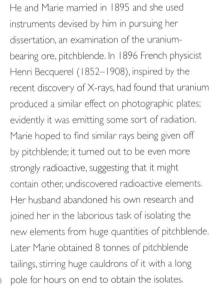

## Hard Times

Marie Curie (1867–1934), née Sklodowska, was the daughter of Polish teachers reduced to straitened circumstances by Russian domination of their homeland. Marie herself struggled to get an education and worked as a governess to help pay for her sister's medical education in Paris. In 1891 she was able to join her sister and studied at the Sorbonne, where she met French chemist Pierre Curie (1859–1906). Pierre had discovered the piezoelectric effect, in which some crystals generate a charge when stressed, and principles governing the magnetic properties of substances.

He and Marie married in 1895 and she used instruments devised by him in pursuing her dissertation, an examination of the uranium-bearing ore, pitchblende. In 1896 French physicist Henri Becquerel (1852–1908), inspired by the recent discovery of X-rays, had found that uranium produced a similar effect on photographic plates; evidently it was emitting some sort of radiation. Marie hoped to find similar rays being given off by pitchblende; it turned out to be even more strongly radioactive, suggesting that it might contain other, undiscovered radioactive elements. Her husband abandoned his own research and joined her in the laborious task of isolating the new elements from huge quantities of pitchblende. Later Marie obtained 8 tonnes of pitchblende tailings, stirring huge cauldrons of it with a long pole for hours on end to obtain the isolates.

• The Nobel committee photos of Marie and Pierre Curie, the first married couple to be jointly honored (though not the last—their daughter and son-in-law were the next).

## Half-life

In July 1898 the Curies announced the discovery of polonium ($_{84}$Po), named for her homeland, and in December the discovery of radium ($_{88}$Ra). They coined the term radioactivity and proved that beta radiation consisted of negatively charged particles, laying the groundwork for the elucidation of atomic structure. In 1903 the Curies shared the Nobel Prize in Physics with Becquerel; three years later Pierre was killed in a traffic accident and Marie took over his teaching post, becoming the first woman to teach at the Sorbonne. During World War I she helped direct medical use of radioactive elements, and in 1916 was awarded the Nobel Prize in Chemistry, for her work on radium, but her health was broken by long exposure to dangerous substances and she died of leukemia. Her daughter Irène (1897–1956), who discovered actinium, went on to win a Nobel prize for creating artificial radioactive elements through neutron bombardment.

The discovery of polonium and radium led others to isolate further radioelements, revealing the entire decay series that led from uranium to lead. This explained why so many radioelements were present in pitchblende: one

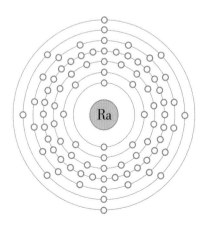

was formed by radioactive decay of the next. Predicting when a single radioactive atom would decay proved impossible, but with a large enough sample it was possible to give a probabilistic answer, stating how long it would take for half the atoms in any given sample to decay. This figure is known as the half-life. For instance, the half-life of radon-222 is 3.8 days. This means that after 3.8 days, half of a sample of Rn-222 will have decayed, and after 7.6 days there will only be a quarter as many atoms.

• The electron configuration of radium (Ra)— an alkaline earth metal element, but also a heavy element, which has a tendency to undergo radioactive decay to radon.

## CARBON DATING

The half-life of the carbon isotope $^{14}$C is 5,730 years, and this can be used to date anything organic. $^{14}$C is created in the upper atmosphere when cosmic radiation bombards carbon dioxide. This means that a small fraction of all the $CO_2$ in the atmosphere contains $^{14}$C, and this is taken in by plants during photosynthesis, and by animals, which eat the plants. As long as the plant/animal is alive it is constantly cycling carbon through its system and will contain a constant proportion of $^{14}$C, but as soon as it dies this proportion begins to drop as the $^{14}$C decays. By measuring the ratio of $^{14}$C to $^{12}$C it is therefore possible to determine when the organism died.

# ORGANIC CHEMISTRY— A VERY BRIEF INTRODUCTION

Organic chemistry is the chemistry of carbon, an element with unique properties that has created an entire sphere of science. As a result of these properties, the chemistry of carbon is also the chemistry of life, and everything that derives from it, from petrol to plastics. Here we give a brief primer of the central concepts and terms.

## The Chemistry of Carbon

A carbon atom has six protons and therefore six electrons; two in its inner shell and four in its outer, valence, shell. These four electrons are the key to carbon's properties, allowing an atom of carbon to form four covalent bonds with other atoms, including other atoms of carbon; these bonds can be single, double or triple. Self-bonding means that carbon can form long chains, and these can act as skeletons to which other elements attach. The number of possible combinations of carbon atoms and their attachments is effectively limitless.

The enormous diversity and complexity of organic chemistry posed a huge challenge for early chemists, once they had begun to draw a distinction between organic and inorganic substances in the late 18th century. Lavoisier showed that the constituents of organic compounds were actually very limited—all included carbon and hydrogen, often with oxygen and occasionally nitrogen. But the further research into organic chemistry progressed, the harder it became to impose any systematic order. At least one pioneer in the field, the great German chemist Justus von Liebig (1803–1873), whose eponymous condenser made it dramatically easier to analyze organic compounds, became so exasperated that he gave up on trying to systematize and turned to applied organic chemistry instead. Not until 1858 was Friedrich August Kekulé (1829–1896) able pull together all the research and formulate a comprehensive theory of chemical structure, emphasizing the importance of carbon backbones or skeletons.

## Hydrocarbons

The simplest organic compounds are created when only hydrogen atoms are attached to this carbon skeleton; these are called "hydrocarbons," and even these are vastly diverse. The naming system for hydrocarbons is based on the type of bonds between atoms in the carbon chain. Molecules with only single bonds are known as alkanes. Molecules with one or more double bonds are called "alkenes," those with one or more triple bonds are called "alkynes," while those in which carbon atoms link up in a ring are called "cyclic hydrocarbons" or "cyclohexenes" (because the rings are made up of six carbon atoms). An important class of cyclohexenes is the aromatics, where the cyclohexene ring has alternating single and double bonds. In an alkane every carbon atom makes four bonds to four different atoms, and these molecules are thus said to be saturated.

There is a distinction between molecular and structural formula for an organic compound, because of the potential for double and triple bonds and for branching of carbon chains. For instance, the hydrocarbon butane has the molecular formula $C_4H_{10}$, but it can adopt one of two forms with different structural formulae. In normal butane the structural formula shows the carbon atoms in a straight chain:

$$CH_3-CH_2-CH_2-CH_3$$

This is a condensed structural formula, different from an expanded one which shows each hydrogen atom separately and all the bonds between each atom. Note that carbon atoms on the end of the chain have three spare bonds and therefore bond to three hydrogen atoms, while those in the middle of the chain use up two bonds linking to the carbon atoms on either side and therefore bond to only two hydrogen atoms.

Another arrangement is possible given this molecular formula, with one carbon atom branching off the chain:

## KEKULÉ'S DREAM

Kekulé famously claimed that the hexagonal ring-shaped structure of the simplest cyclohexene, benzene ($C_6H_6$), came to him in a dream. He explained that, musing on the problem of benzene's structure, "I turned my chair to the fire and dozed. Again the atoms were gambolling before my eyes … all twining and twisting in snakelike motion. But look! What was that? One of the snakes had seized hold of its own tail, and the form whirled mockingly before my eyes. As if by a flash of lightning I awoke …"

$$CH_3-CH-CH_3$$
$$|$$
$$CH_3$$

A compound with the same molecular formula but different structural formula is known as an isomer, so this is known as isobutane.

When an element other than hydrogen bonds with an organic molecule it is known as a functional group. Important functional groups include the alcohols, where an $-OH$ group is bound to the carbon backbone; and the amines, where a nitrogen containing functional group, $-NH_2$, is involved. The simplest form of alcohol is methanol (aka methyl or wood alcohol): $CH_3OH$. Ethanol, the alcohol found in wine, beer and spirits, is $CH_3CH_2OH$.

# INDEX

# TERMS

**Acid**  compound that raises concentration of $H^+$ ions in water.

**Activation energy**  energy boost that gets a reaction started.

**Alkali**  base that dissolves in water to produce hydroxide ions.

**Atomic mass**  mass of an atom in atomic mass units; 1 amu = ¹/₁₂th the mass of an atom of carbon-12.

**Atomic weight**  average mass per atom of naturally occurring forms of an element, weighted to take account of each isotope's fractional abundance.

**Base**  compound that reacts with an acid to produce a salt.

**Combustion**  reaction where a compound combines with oxygen; otherwise known as burning redox, short for reduction-oxidation.

**Covalent bond**  bond formed when two atoms share a pair of electrons; the electrons effectively take up a new orbit encompassing both atoms.

**Endothermic**  reaction that absorbs heat energy from the surroundings.

**Exothermic**  reaction that generates heat.

**Inorganic chemistry**  chemistry of the elements and the compounds of elements other than carbon.

**Ion**  atom that has lost or gained one or more electrons to become positively or negatively charged.

**Isotope**  atoms of the same element with the same atomic number but with different numbers of neutrons and therefore different mass number. They have the same number of protons in the nucleus of each atom.

**Kinetic energy**  energy of motion that particles have, and determines the speed and force of their motion.

**Mass number**  combined total of protons and neutrons in the nucleus of an atom.

**Organic chemistry**  chemistry of compounds of carbon.

**Oxidation**  loss of electrons; chemical reaction where electrons are lost, as when reacting with oxygen.

**pH scale**  logarithmic scale of acidity/alkalinity expressing concentration of $H^+$ ions.

**Pneumatic chemistry**  study of gases.

**Radioactivity**  decomposition or decay of an unstable nucleus, involving the loss and/or transformation of subatomic particles with release of energy.

**Valency**  combining power of an atom, ion, or radical; number of hydrogen atoms it can bond with.